# CONVEX ANALYSIS

# TEXTBOOKS in MATHEMATICS

## Series Editors: Al Boggess and Ken Rosen

PUBLISHED TITLES

RISK ANALYSIS IN ENGINEERING AND ECONOMICS, SECOND EDITION
Bilal M. Ayyub

COUNTEREXAMPLES: FROM ELEMENTARY CALCULUS TO THE BEGINNINGS OF ANALYSIS
Andrei Bourchtein and Ludmila Bourchtein

INTRODUCTION TO THE CALCULUS OF VARIATIONS AND CONTROL WITH MODERN APPLICATIONS
John T. Burns

MIMETIC DISCRETIZATION METHODS
Jose E. Castillo

AN INTRODUCTION TO PARTIAL DIFFERENTIAL EQUATIONS WITH MATLAB®, SECOND EDITION
Mathew Coleman

RISK MANAGEMENT AND SIMULATION
Aparna Gupta

ABSTRACT ALGEBRA: AN INQUIRY-BASED APPROACH
Jonathan K. Hodge, Steven Schlicker, and Ted Sundstrom

QUADRACTIC IRRATIONALS: AN INTRODUCTION TO CLASSICAL NUMBER THEORY
Franz Holter-Koch

GROUP INVERSES OF M-MATRICES AND THEIR APPLICATIONS
Stephen J. Kirkland

AN INTRODUCTION TO NUMBER THEORY WITH CRYPTOGRAPHY
James Kraft and Larry Washington

CONVEX ANALYSIS
Steven G. Krantz

DIFFERENTIAL EQUATIONS: THEORY, TECHNIQUE, AND PRACTICE, SECOND EDITION
Steven G. Krantz

ELEMENTS OF ADVANCED MATHEMATICS, THIRD EDITION
Steven G. Krantz

REAL ANALYSIS AND FOUNDATIONS, THIRD EDITION
Steven G. Krantz

PUBLISHED TITLES CONTINUED

APPLYING ANALYTICS: A PRACTICAL APPROACH
Evan S. Levine

ADVANCED LINEAR ALGEBRA
Nicholas Loehr

DIFFERENTIAL EQUATIONS WITH MATLAB®: EXPLORATION, APPLICATIONS, AND THEORY
Mark A. McKibben and Micah D. Webster

APPLICATIONS OF COMBINATORIAL MATRIX THEORY TO LAPLACIAN MATRICES OF GRAPHS
Jason J. Molitierno

ABSTRACT ALGEBRA: AN INTERACTIVE APPROACH
William Paulsen

ADVANCED CALCULUS: THEORY AND PRACTICE
John Srdjan Petrovic

COMPUTATIONS OF IMPROPER REIMANN INTEGRALS
Ioannis Roussos

TEXTBOOKS in MATHEMATICS

# CONVEX ANALYSIS


Steven G. Krantz

Washington University

St. Louis, Missouri, USA

CRC Press is an imprint of the

Taylor & Francis Group an **informa** business

A CHAPMAN & HALL BOOK

CRC Press
Taylor & Francis Group
6000 Broken Sound Parkway NW, Suite 300
Boca Raton, FL 33487-2742



Printed on acid-free paper
Version Date: 20140828

International Standard Book Number-13: 978-1-4987-0637-7 (Paperback)



**Visit the Taylor & Francis Web site at**
**http://www.taylorandfrancis.com**

**and the CRC Press Web site at**
**http://www.crcpress.com**

With thoughts of Archimedes.

# Table of Contents

# Preface

Convexity is an old idea. It even occurs, in some vestigial form, in Archimedes's treatment of arc length. The concept appeared intermittently in the work of Fermat, Cauchy, Minkowski, and others. Even Johannes Kepler treated convexity. The first specific definition of convexity was given by Herman Minkowski in 1896. Convex *functions* were introduced by Jensen in 1905—see [JEN]. But it can be said that the subject as such was not really formalized until the seminal 1934 tract *Theorie der konvexen Körper* of Bonneson and Fenchel.

Today convex geometry is a mathematical subject in its own right. Classically oriented treatments, like that of Frederick Valentine, work from the elementary definition that a domain $\Omega$ in the plane or in $\mathbb{R}^N$ is convex if, whenever $P, Q \in \Omega$, then the segment $\overline{PQ}$ connecting $P$ to $Q$ also lies in $\Omega$.

In fact this very simple idea gives forth a very rich theory. But it is not a theory that interacts naturally with mathematical analysis. For analysis, one would like a way to think about convexity that is expressed in the language of functions, and that involves perhaps derivatives. In point of fact it is very natural to formulate convexity at the boundary of a domain in terms of the second fundamental form of Riemannian geometry. Alternatively, one may study convexity in terms of the Hessian of an exhaustion function for the domain. We discuss and define these concepts in Chapter 2.

The approach described in the last paragraph in fact is very rich and productive. It rather naturally subsumes the more classical approach presented in the paragraph before, and provides a new set of tools for analytic study. It also has a symbiotic relationship with partial differential equations, harmonic analysis, control theory, and many other parts of mathematics.

Our goal in this book is to present and to study this more analytic way to think about convexity theory. Although this means of doing things is well known to the experts, it is not well documented in the literature. There is no existing book that develops this point of view.

One of the nice features of the analytic way of looking at convexity is that it is entirely self-contained. The reader with only basic background in real analysis (and perhaps a little linear algebra) can get a lot out of reading this book. We will also provide some background in the classical, geometric theory. Thus the reader will get a glimpse of how modern mathematics is developed, and of how geometric ideas may be studied analytically.

The book has copious examples and plenty of figures. After all, this is still a geometric subject and the student will do well to think of questions geometrically. Plenty of applications will be provided, both to geometry and to analysis. Each chapter ends with a set of exercises.

Even though convexity is a fairly basic topic, it is unavoidable that this book must sometimes draw on more advanced ideas. To make things easier, we have provided an Appendix called Technical Tools that helps to acquaint the reader with some of these more sophisticated topics.

There are several modern works on convexity that arise from studies of functional analysis. Those studies are certainly not disjoint from what we are doing here. But they live in a much more abstract context (i.e., infinite dimensions) that is ill-suited to our more basic purposes. We confine our attention to finite-dimensional Euclidean space, and sometimes even to the Euclidean plane. The consequence is a rich and tactile theory that is rewarding to study and rich in results.

The book ends with a detailed Table of Notation and a rather thorough Glossary to help the reader with unfamiliar terms. We want this book to be as user-friendly as feasible, given the broad range of material that we endeavor to cover.

It is a pleasure to thank Fernando Gouvea and the Carus monograph committee for exceptionally careful readings of my many manuscripts and for innumerable constructive suggestions.

We trust that those interested in both geometry and analysis will get much from this book, and will be anxious to turn to the literature for more substance and depth after finishing this volume.

St. Louis, Missouri

Steven G. Krantz

# Biography of Steven G. Krantz

Steven G. Krantz was born in San Francisco, California in 1951. He received the B.A. degree from the University of California at Santa Cruz in 1971 and the Ph.D. from Princeton University in 1974.

Krantz has taught at UCLA, Princeton University, Penn State, and Washington University in St. Louis. He was chair of the latter department for five years.

Krantz has had 9 Masters students and 18 Ph.D. students. He has written more than 75 books and more than 195 scholarly papers. He edits 6 journals, and is managing editor of 4. He is the founding editor of *The Journal of Geometric Analysis* and *Complex Analysis and its Synergies.*

Krantz has won the Chauvenet Prize, the Beckenbach Book Award, and the Kemper Prize. He was recently named to the Sequoia High School Hall of Fame. He is an AMS Fellow.

# Chapter 0

# Why Convexity?

**Prologue:** In this brief chapter we motivate the study of convex sets and convex functions with a little bit of history and a little bit of philosophy.

Convex functions are an important device for the study of extremal problems. We shall develop that theme in the present book. They are also important analytic tools. The fact that a convex function can have at most one minimum and no maxima is a notable piece of information that proves to be quite useful.

A convex function is also characterized by the nonnegativity of its second derivative. This is useful information that interacts nicely with the ideas of calculus.

The study of convex functions (and sets) is rich and varied. This book will acquaint you with its many aspects.

The idea of convexity finds its roots in the ideas of the ancient Greeks. What is it about convexity that is compelling, that makes us want to study the concept?

First of all, it is an elementary geometric idea that has immediate appeal. Second, it is based on positivity (at least as we present it here), and positivity is one of the most powerful ideas in mathematical analysis. Third, it is a concept that is stable under linear transformations and other elementary operations. Fourth, and perhaps most important, it is a notion of nonlinearity that is manageable and understandable.

The idea of convex set is quite old, but the idea of convex function is relatively new. This is true in part because the concept of function is rather new. The modern, rigorous definition of function was only formulated by

Goursat in the 1920s. The present book places a great emphasis on convex functions.

And why is that? Sets are not objects that are easily manipulated. There are no algebraic operations on sets. But functions have a great deal of algebraic structure built into them. We can add, subtract, multiply, divide, and compose functions. We can also differentiate them and integrate them. Thus a great deal of modern mathematics is built on functions. And there is a lot to say about convex functions.

As we shall see, convex functions are characterized by the property that their second derivatives are positive. Of course a positivity condition like this is preserved under many different operations—including sum, supremum, and modulus. Products of convex sets are convex, and so are intersections. There is a great lore of convex sets and functions, and many appealing theorems.

It is fascinating that convex sets and functions can be described in such simple language, yet their theory is substantial. Convex functions play a major role in optimization theory, and in von Neumann's minimax theorem. They are an essential part of any modern mathematics education. Yet they are poorly represented in the literature. Any modern course in mathematical analysis ought to treat convex functions, yet usually it does not. Convex functions should be part of our *lingua franca*, but they are often shunted to the side.

Why do convex functions and sets get less attention than they deserve? The answer is partly that it is such a classical subject; it does not "feel" modern. It is not a topic that Cauchy and Riemann and Weierstrass dwelled on, so why should we spend our time with it? Yet convex functions are an essential part of optimization theory and should certainly command our interest. Convex functions should be part of every analyst's toolkit, and all mathematicians should be aware of them.

It is the goal of this book to present a modern, yet streamlined, introduction to the theory of convexity. We want to touch on the key ideas, yet not get bogged down in technicalities. We want to show the beauty of the subject, and to hit the high points in a compelling fashion. The proof is in the pudding, and we shall begin at the beginning.

# Chapter 1

# Basic Ideas

**Prologue:** This chapter provides background material from the very classical theory of convex sets. Many of the ideas here are one hundred years old or more. You will see no analysis in this chapter—only elementary geometry.

Yet there are fascinating results in this elementary context: the theorems of Helly, Kirchberger, Carathéodory, and others are nontrivial and useful. And we will make good use of them later in the book when we are doing analysis.

The ideas about approximating an arbitrary convex set by a more regular convex set (like a polygon or a simplex) are also useful and nontrivial. They round out the chapter and set us up for our analytic studies that begin in Chapter 2.

## 1.0 Introduction

Convexity is an old subject in mathematics. The concept appeared intermittently through the centuries, but the subject was not really formalized until the seminal tract of Bonneson and Fenchel [BOF]. See also [FEN] for the history. Modern treatments of convexity may be found in [LAY] and [VAL]. See also [BOV], [HOR], [LAU], [ROC], [SIM], and [TUY].

In what follows, we let the term *domain* denote a connected, open set. Figure 1.1 shows two domains, and Figure 1.2 shows two non-domains. We usually denote a domain by $\Omega$. If $\Omega$ is a domain and $P, Q \in \Omega$ then the *closed segment* determined by $P$ and $Q$ is the set

$$\overline{PQ} \equiv \{(1-t)P + tQ : 0 \leq t \leq 1\}.$$

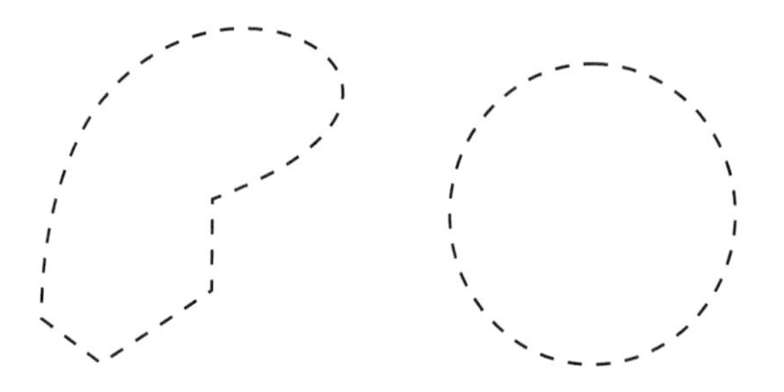


Figure 1.1: Illustration of domains.

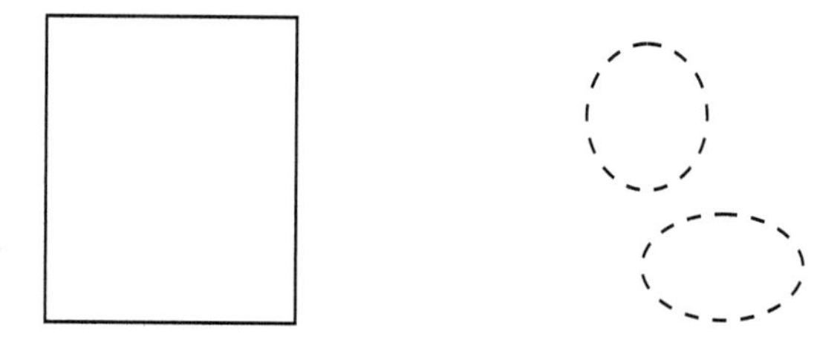


Figure 1.2: Illustration of non-domains.

See Figure 1.3.

Most of the classical treatments of convexity rely on the following synthetic definition of the concept:

**Definition 1.1** Let $S \subseteq \mathbb{R}^N$ be a set. We say that $S$ is *convex* if, for $P, Q \in S$, the closed segment $\overline{PQ}$ from $P$ to $Q$ lies in $S$. See Figures 1.4 and 1.5.

In what follows, we shall refer to the geometric property described in this definition as *geometric convexity*. This will be in contrast to the notion of analytic convexity (introduced in Chapter 2), which will be the main focus of the present book.

Works such as [LAY] and [VAL] treat theorems of Helly and Kirchberger—about configurations of convex sets in the plane, and points in those convex sets. However, studies in analysis and differential geometry (as opposed to synthetic geometry) require results—and definitions—of a different type. We need hard analytic facts about the shape of the boundary—formulated in

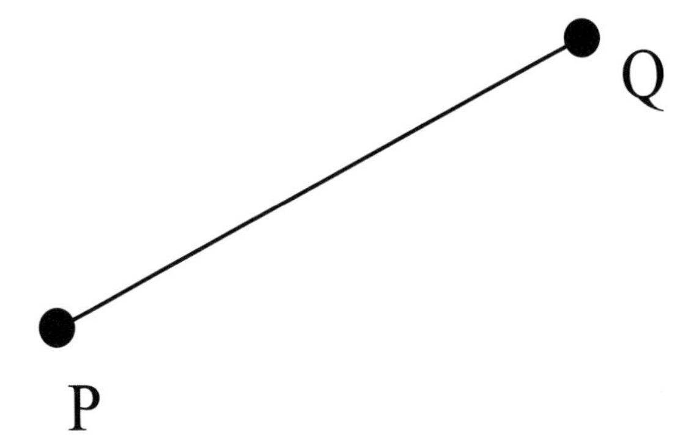


Figure 1.3: A segment.

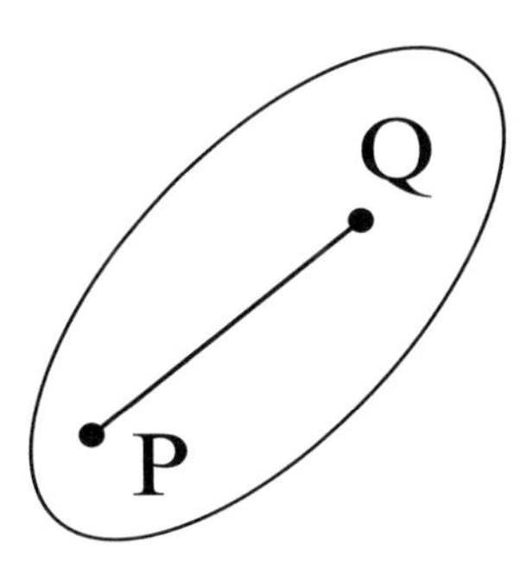


Figure 1.4: The classical definition of convexity.

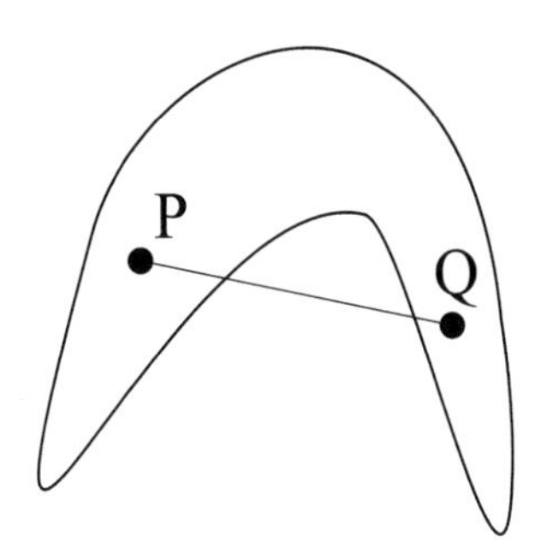


Figure 1.5: A non-convex set.

differential-geometric language. We need invariants that we can calculate and estimate. In particular, it is useful to have a definition of "convex" that is local, that can be measured point by point. The classical definition of convexity—using segments—treats the whole domain at once. That is too imprecise, too "soft," and it is *not* quantitive. The differential-geometric view of the world is that we want to measure things point by point. And that is the perspective that we wish to explore in the present book.

## 1.1 A Deeper Look at the Classical Theory

For context and background, we now treat some of the great classical theorems in the subject of convexity theory. Some of these date back to the early twentieth century. We stress once again that these theorems do *not* represent the point of view that we are promoting in the present book. Rather, they provide a more gentle introduction to convexity theory, and provide motivation for the more demanding material on analytic convexity that will follow. We shall use some of these classical results as the book develops.

The first result of Radon is mostly presented by way of background. It has intrinsic interest, but is significant for us because it is used in subsequent proofs.

First we need a definition.

**Definition 1.2** Let $A \subseteq \mathbb{R}^N$ be any set. The *convex hull* of $A$ is the intersection of all geometrically convex sets that contain $A$. The *closed convex hull* of $A$ is the intersection of all closed, geometrically convex sets that contain $A$.

This concept bears some discussion. If $A \subseteq \mathbb{R}^N$ is *any* set, and if $P, Q \in A$, then any point of the form $(1-t)P + tQ$ for $0 \le t \le 1$ will lie in the convex hull of $A$—just by the classical definition of geometric convexity. But we may in fact iterate this statement to derive the noteworthy assertion that, if $P_1, P_2, \ldots, P_k \in A$, then

$$a_1 P_1 + a_2 P_2 + \cdots + a_k P_k \tag{1.3}$$

lies in the convex hull for any nonnegative numbers $a_1, a_2, \ldots, a_k$ with $a_1 + a_2 + \cdots + a_k = 1$. The expression (1.3) is a standard device for calculating the convex hull. And in fact it is very natural to use (1.3) to provide an alternative definition of "convex hull."

**Example 1.4** Let $A \subseteq \mathbb{R}^2$ be given by

$$A = \{(x,0) : -1 \le x \le 1\} \cup \{(0,y) : -1 \le y \le 1\} .$$

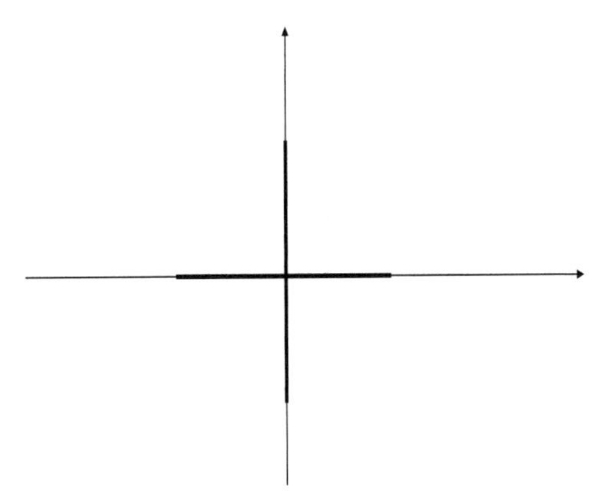

Figure 1.6: A cross in the plane.

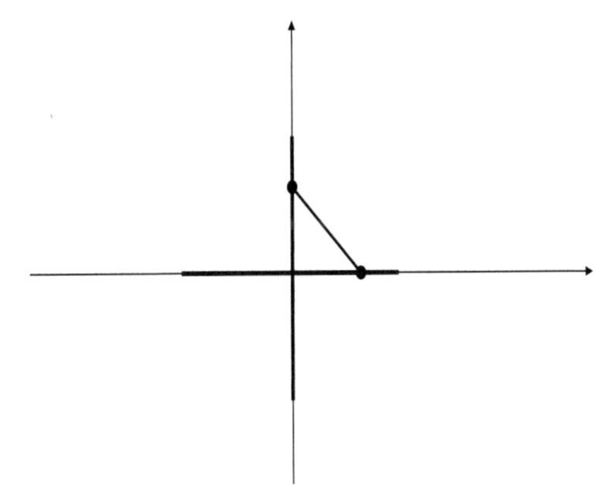

Figure 1.7: A segment in the convex hull.

So $A$ is a cross in the plane. See Figure 1.6.

If $(x, 0), (0, y) \in A$ then, according to our preceding observations, $a(x, 0) + (1 - a)(0, y)$ lies in the convex hull for any $0 \leq a \leq 1$. Such a segment is shown in Figure 1.7.

More generally, if $p_1, p_2, \ldots, p_k$ are points of $A$, and if $a_1, a_2, \ldots, a_k$ are nonnegative numbers with $a_1 + a_2 + \cdots + a_k = 1$, then $a_1 p_1 + a_2 p_2 + \cdots + a_k p_k$ lies in the convex hull of $A$. Figure 1.8 illustrates this idea. ∎

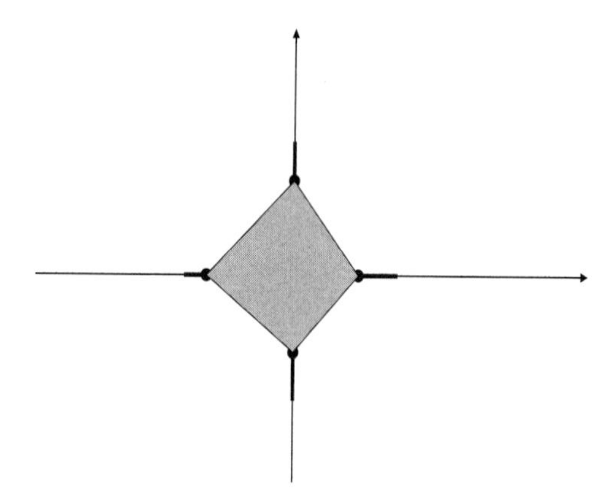

Figure 1.8: The convex hull in general.

PROPOSITION 1.5 (RADON) *Let $Q = \{p_1, p_2, \ldots, p_{N+2}\}$ be a finite set of distinct points in $\mathbb{R}^N$ with $N \geq 1$. Then $Q$ can be partitioned into two distinct sets $\mathcal{P}, \mathcal{N}$ whose convex hulls intersect.*

**Proof:** Consider the system of equations

$$\begin{aligned} \sum_{j=1}^{N+2} a_j p_j &= 0 \\ \sum_{j=1}^{N+2} a_j &= 0\,. \end{aligned}$$

The first of these is actually $N$ equations and the second is one equation. So we have $N+1$ equations in $N+2$ unknowns. Thus we may be sure that there exists a nontrivial solution set $\{a_1, a_2, \ldots, a_{N+2}\}$.

Let $\mathcal{P}$ consist of those points $p_j$ such that $a_j$ is positive and let $\mathcal{N}$ consist of those points $p_j$ such that $a_j$ is negative or zero. We claim that $\mathcal{P} \cup \mathcal{N}$ is the partition of $Q$ into two subsets with intersecting convex hulls.

To see this, let

$$A = \sum_{j \in P} a_j = -\sum_{j \in N} a_j\,.$$

Define

$$p = \sum_{j \in P} \frac{a_j}{A} \cdot p_j = \sum_{j \in N} \frac{-a_j}{A} \cdot p_j\,.$$

This last equation shows $p$ to be both in the convex hull of $\mathcal{P}$ and in the convex hull of $\mathcal{N}$. That proves the result. Figure 1.9 illustrates Radon's theorem.

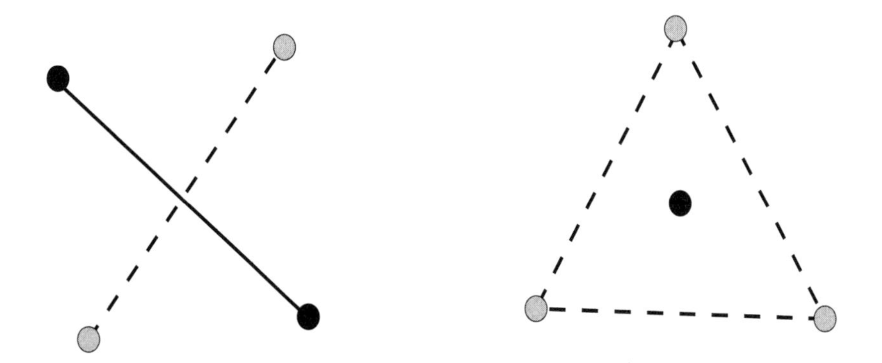

Figure 1.9: Two sets of four points each of which illustrate Radon's theorem. Each picture shows the partition (into black and grey points) that satisfy the conclusion.

□

Now one of the big results of classical convexity theory is the following.

THEOREM 1.6 (HELLY) *Let $X_1, X_2, \ldots, X_k$ be a finite collection of convex subsets of $\mathbb{R}^N$, where $k > N$. If the intersection of every $N+1$ of these sets is nonempty, then the entire collection has nonempty intersection. That is to say, we conclude that*

$$\bigcap_{j=1}^{k} X_j \neq \emptyset .$$

*See Figure 1.10.*

**Remark 1.7** This theorem is reminiscent of a basic result from the theory of compact sets. Namely, if $\{S_\alpha\}$ are compact, and if every finite subcollection has nonempty intersection, then the entire collection has nonempty intersection. Some proofs of Helly's theorem actually use this topological result.

**Proof of Helly's Theorem:** First suppose that $k = N+2$. By hypothesis, there is a point $x_1$ that lies in $X_2 \cap X_3 \cap \cdots \cap X_k$. Likewise, for each $j = 2, 3, \ldots, k$, there is a point $x_j$ that is common to all $X_\ell$ except possibly $X_j$. Now we apply Radon's theorem to the set $Q = \{x_1, x_2, \ldots, x_k\}$. So there are subsets $\mathcal{P}$ and $\mathcal{N}$ that partition $Q$ so that the convex hull of $\mathcal{P}$ intersects the convex hull of $\mathcal{N}$. Let $p$ be a point of that intersection. We claim that $p \in \cap_{\ell=1}^{k} X_j$.

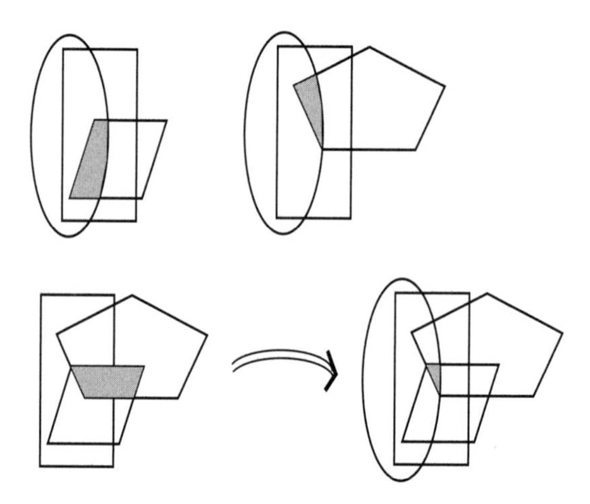

Figure 1.10: Helly's theorem.

To see this, take $j \in \{1, 2, \ldots, k\}$. Notice that the only element of $Q$ that may not be in $X_j$ is $x_j$. If $x_j \in \mathcal{P}$, then $x_j \notin \mathcal{N}$, so $X_j \supseteq \mathcal{N}$. Since $X_j$ is convex, we conclude that $X_j$ contains the convex hull of $\mathcal{N}$, and hence $p \in X_j$. Likewise, if $x_n \notin \mathcal{P}$, then $X_j \supseteq \mathcal{P}$. So the same reasoning shows that $p \in X_j$. Since $p$ is thus in every $X_j$, it must also be in the intersection.

Now suppose inductively that $k > N + 2$ and that the statement is true for $k - 1$. The preceding argument shows that any subcollection of $N + 2$ sets will have nonempty intersection. Now we consider the collection of sets where we replace $X_{k-1}$ and $X_k$ with $X_{k-1} \cap X_k$. In this new collection, every subcollection of $N + 1$ sets will have nonempty intersection. The inductive hypothesis therefore applies, and we may conclude that this new collection has nonempty intersection. Therefore so does the original collection. That proves the result. □

We next treat a cornerstone of classical convexity theory that is due to Kirchberger. We first need an ancillary result of Carathéodory.

LEMMA 1.8 (CARATHÉODORY) *Let $S \subseteq \mathbb{R}^N$ be any set and $x \in \mathbb{R}^N$. Assume that $x$ lies in the convex hull of $S$. Then there is a subset $T$ of $S$, consisting of at most $N + 1$ points, so that $x$ lies in the convex hull of $T$.*

Carathéodory's result is intuitively appealing. Figure 1.11 illustrates the idea.

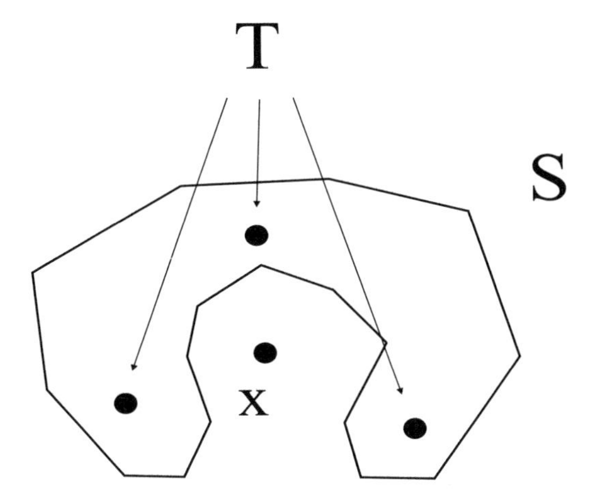


Figure 1.11: Carathódory's theorem.

**Proof of Carathéodory's Lemma:** Let $x$ be a point in the convex hull of $S$. Then certainly $x$ is a convex combination of a finite number of points in $S$:

$$x = \sum_{j=1}^{k} \lambda_j x^j .$$

Here each $x^j$ lies in $S$, every $\lambda_j$ is positive, and $\sum_{j=1}^{k} \lambda_j = 1$.

We suppose now that $k > N + 1$. Then the vectors $x^2 - x^1$, $x^3 - x^1$, ..., $x^k - x^1$ are linearly independent. Hence there are real scalars $\mu_2, \mu_3, \ldots, \mu_k$, not all zero, so that

$$\sum_{j=2}^{k} \mu_j (x^j - x^1) = 0 .$$

If $\mu_1$ is defined to be

$$\mu_1 \equiv - \sum_{j=2}^{k} \mu_j ,$$

then

$$\begin{aligned} \sum_{j=1}^{k} \mu_j x^j &= 0 \\ \sum_{j=1}^{k} \mu_j &= 0 . \end{aligned}$$

Also not all of the $\mu_j$ are equal to 0. Thus we may suppose that at least one of the $\mu_j$ is positive.

Thus we may write

$$x = \sum_{j=1}^{k} \lambda x^j - \alpha \sum_{j=1}^{k} \mu_j x^j = \sum_{j=1}^{k} (\lambda_j - \alpha\mu_j) x^j$$

for any real $\alpha$. In particular, the equality will hold if $\alpha$ is defined by

$$\alpha \equiv \min_{1 \le j \le k} \left\{ \frac{\lambda_j}{\mu_j} : \mu_j > 0 \right\} = \frac{\lambda_i}{\mu_i} .$$

Here $i$ is a particular index.

Note that $\alpha > 0$ and that, for every $j$ between 1 and $k$,

$$\lambda_j - \alpha\mu_j \ge 0 .$$

In particular, $\lambda_i - \alpha\mu_i = 0$ by the very definition of $\alpha$. Thus

$$x = \sum_{j=1}^{k} (\lambda_j - \alpha\mu_j) x^j ,$$

where every $\lambda_j \ \alpha\mu_j$ is nonnegative, their sum is 1, and also $\lambda_i - \alpha\mu_i = 0$. The point is that $x$ is represented as a convex combination of at most $k-1$ points of $S$. This process can be repeated until $x$ is represented as a convex combination of at most $N+1$ points of $S$.

That completes the proof. □

We also note—again see [VAL]—that it is a standard result that two disjoint compact, convex sets in space can be separated by a hyperplane.[1] See the treatment in the next section, and also Figure 1.12.

THEOREM 1.9 (KIRCHBERGER) *Let $S$ and $T$ be two finite sets in $\mathbb{R}^N$. Assume that $k + \ell \le N$ and that any $k+1$ points of $S$ may be separated from any $\ell+1$ points of $T$ by a hyperplane in $\mathbb{R}^N$. Then $S$ may be separated from $T$ by a hyperplane of $\mathbb{R}^N$. See Figure 1.13.*

If one consults a classical text like [VAL], or even a modern functional analysis text (see also [SIM]), he/she will see that the idea of separating convex sets with hyperplanes is one of the most basic and important ideas in the

[1] A hyperplane in $\mathbb{R}^N$ is an $(N-1)$-dimensional plane, that is to say, a plane of codimension 1. See the Appendix.

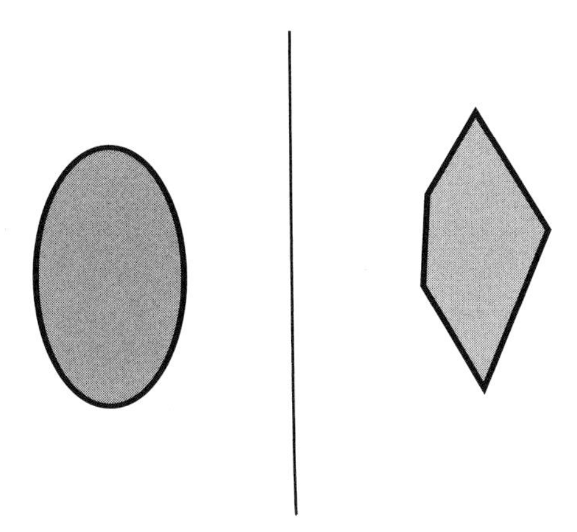

Figure 1.12: Separation of two convex sets by a hyperplane.

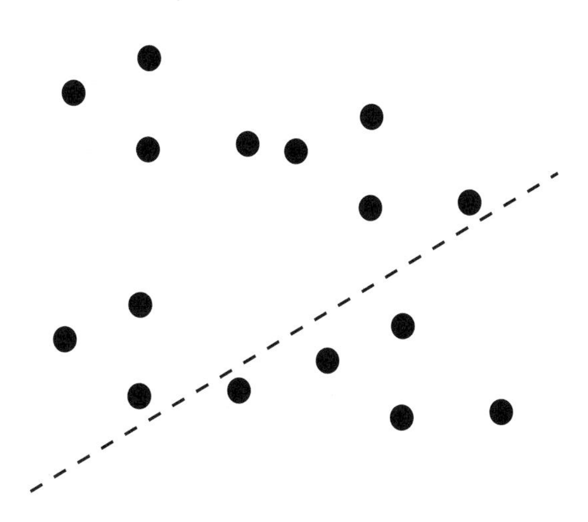

Figure 1.13: Kirchberger's theorem.

subject. We treat it in detail momentarily.

**Proof of Kirchberger's Theorem:**

This proof comes from [SHI].

We begin by making the apparently stronger assumption that the hypothesis of the theorem holds for $k \leq N$ and $\ell \leq N$.

By the remarks preceding the statement of the theorem, we see that it suffices for us to prove the following:

> The convex hulls $H(S)$ and $H(T)$ have no points in common. (1.9.1)

If (1.9.1) were not true, then (by Carathéodory's result) any point of $H(S) \cap H(T)$ would belong to two simplices of dimensions not exceeding $N$ with vertices in $S$ and $T$ respectively. But our hypothesis tells us that the vertices of the one simplex may be separated from those of the other simplex by a hyperplane of $\mathbb{R}^N$, and thus their convex hulls may be similarly separated. That contradiction establishes our assertion.

Now we show that, if there are two subsets of $S$, $T$ consisting of $k+1$, $\ell+1$ points respectively—with $k \leq N$ and $\ell \leq N$—which cannot be separated by a hyperplane, then there must exist two subsets consisting of $k_1+1$, $\ell_1+1$ points respectively of $S$, $T$, with $k_1 + \ell_1 \leq N$ and which also cannot be separated by a hyperplane.

In point of fact, let $A$, $B$ be two sets in $\mathbb{R}^N$, consisting of $k+1$ and $\ell+1$ points respectively (here $k \leq N$ and $\ell \leq N$), which cannot be separated by hyperplanes. Thus the set $H(A0 \cap H(B) \neq \emptyset$. Let $p_0$ be a point of this intersection. Let $\sigma^{k'}$ and $\tau^{\ell'}$ be simplices (the superscripts represent the vertex sets) with vertices in $A$ and $B$ respectively that contain $p_0$ in their relative interiors. Denote by $E^{k'}$, $F^{\ell'}$ the "minimal flats" (that is to say, the linear surfaces of minimum dimension) containing $\sigma^{k'}$ and $\tau^{\ell'}$ respectively. In case $k' + \ell' > N$, the flat $E^{k'} \cap F^{\ell'}$ is at least $\ell$-dimensional and hence contains a ray issuing from $p_0$. If $p_1$ is the first point on that ray which does not belong to the relative interior of both $\sigma^{k'}$ and $\tau^{\ell'}$, then $p_1$ is an interior point of some faces (at least one of these is proper) $\sigma^{k''}$ and $\tau^{\ell''}$ of $\sigma^{k'}$ and $\tau^{\ell'}$ respectively. Here $k'' + \ell'' < k' + \ell'$. By repeating this process of reducing the dimension-sum, we can ultimately make the sum smaller than or equal to $N$, which yields the required result. That is the theorem. □

The unifying theme of the classical theorems we have discussed here (Radon, Helly, and Kirchberger) is one of robustness. Geometrically convex sets have

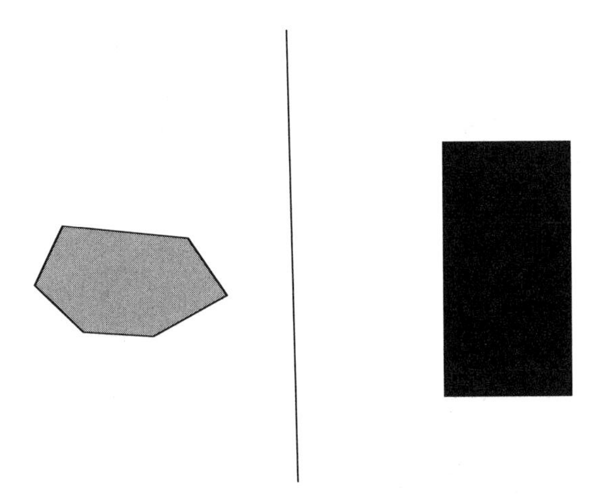

Figure 1.14: Separation of a closed convex set from a compact convex set.

striking properties regarding intersection, and the theorems make those explicit and, at least in some cases, quantitative. In the next section we begin to treat separation properties, which certainly distinguish convex sets from arbitrary sets.

## 1.2 Separation Theorems

In this brief section we treat some separation theorems for convex sets. The first of these is particularly simple.

THEOREM 1.10 *Let $A$ be a closed convex set and $B$ be a compact convex set, both in $\mathbb{R}^N$. Assume that these sets are disjoint. Then there is a hyperplane in $\mathbb{R}^N$ that separates $A$ from $B$. More precisely, there is a hyperplane $\mathcal{H}$ such that $A$ lies on one side of $\mathcal{H}$ and $B$ lies on the other side of $\mathcal{H}$. See Figure 1.14.*

**Proof:** It is a standard fact from analysis/topology that there is a point $a \in A$ and a point $b \in B$ such that $|a - b|$ minimizes the distance of points in $A$ from points in $B$. Here $a$ and $b$ are of course distinct and $|a - b| > 0$.

Let $p$ be the midpoint of the segment $\overline{ab}$ and let $\mathcal{H}$ be the hyperplane through $p$ that is orthogonal to $\overline{ab}$. Then $\mathcal{H}$ is the hyperplane that we seek. $\square$

A variant of the result is as follows.

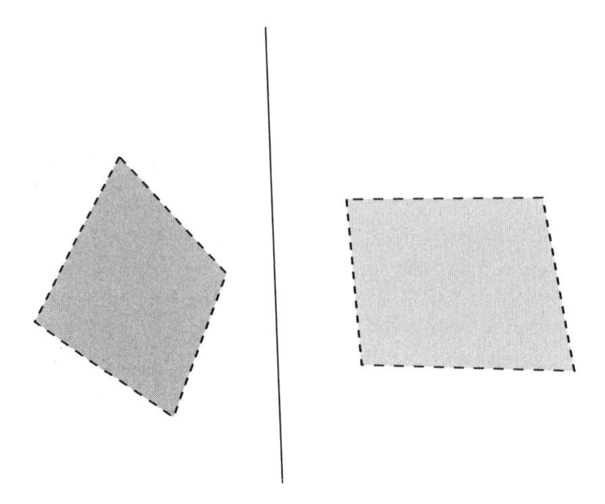

Figure 1.15: Separation of open convex sets.

THEOREM 1.11 *Let $A$ and $B$ be disjoint, open convex sets in $\mathbb{R}^N$. Then there is a hyperplane $\mathcal{H}$ that separates $A$ from $B$. See Figure 1.15.*

**Proof:** Now $A$ can be exhausted by an increasing union of compact, convex sets $A_j$, and $B$ can be exhausted by an increasing union of compact, convex sets $B_j$. The first version of our theorem then provides, for each $j$, a separating hyperplane $\mathcal{H}_j$. Each such hyperplane has a unit normal vector $\nu_j$ and a base point $p_j$. Of course the sequence of normal vectors can be thought of as a subset of the unit sphere, so it will have a convergent subsequence. And then there is a sub-subsequence so that the base points converge (because they can all be taken to lie in a large, bounded, closed ball). So we get a limit normal vector and a limit base point. That, of course, gives a limit hyperplane $\mathcal{H}_0$. This hyperplane certainly separates $A$ from $B$. □

## 1.3 Approximation of Convex Sets

It is quite common in the subject of convex geometry to first prove a result for convex polyhedra, or for smoothly bounded strongly analytically convex domains, and then to do an approximation argument. In the present section we lay some of the foundations for that sort of analysis.

PROPOSITION 1.12 *Let $K \subseteq \mathbb{R}^N$ be a compact, convex set. Let $U$ be an open neighborhood of $K$. Then there exist $K_1$, $K_2$ such that*

1. *Each $K_j$ is compact and convex;*

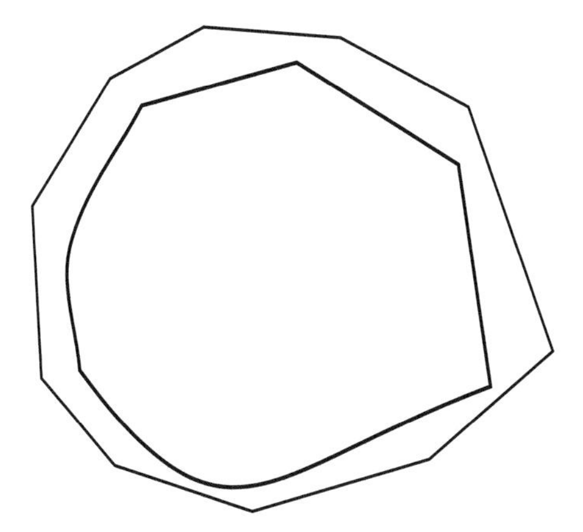

Figure 1.16: Approximation of a convex set by a polyhedron.

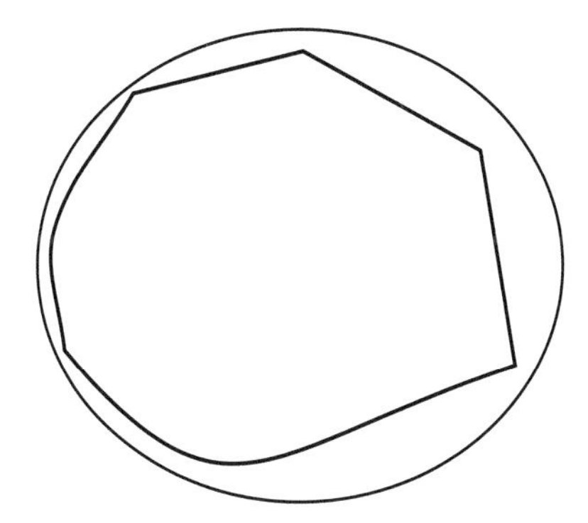

Figure 1.17: Approximation of a convex set by a smoothly bounded, strongly analytically convex domain.

2. *We have* $K \subset K_j \subset U$ *for each* $j$*;*

3. *The set* $K_1$ *is a polyhedron;*

4. *The set* $K_2$ *is smoothly bounded and strongly analytically convex (all boundary curvatures positive). See Figures 1.16 and 1.17.*

**Proof of the Proposition:** We may assume that $U$ is convex.

Each point of $K$ lies in the interior of a simplex that in turn lies in $U$ (see Exercise 8 at the end of the chapter for more on simplices). By compactness, we may cover $K$ with finitely many of these simplices. Then the convex hull of the union of that collection of simplices will serve as $K_1$.

For our construction of $K_2$, we may assume that the origin lies in $K$. Using the Minkowski functional as our notion of distance, and denoting it by $p$, we may think of

$$K = \{x \in \mathbb{R}^N : p(x) \leq 1\}.$$

Now let $\chi$ be a nonnegative $C_c^\infty$ function, supported in the unit ball, with integral equal to 1. Now define

$$p_\epsilon(x) = \int p(x - \epsilon t)\chi(t)\,dt + \epsilon|x|^2.$$

We see that $p_\epsilon$ is strictly convex. Also, for $\epsilon > 0$ small,

$$p_\epsilon(x) \leq p(x) + C_1 x \leq 1 + C_1\epsilon\,, \quad \text{all } x \in K.$$

Also

$$p_\epsilon(x) \geq C_2 > 1 \quad \text{for} \quad x \notin U.$$

Now, for small $\epsilon$, the unique minimum point of $p_\epsilon$ belongs to $K$, hence the domain

$$K_2 \equiv \{x : p_\epsilon(x) \leq 1 + 2C_1\epsilon\}$$

has $C^\infty$ boundary with strictly positive principal curvatures since $p_\epsilon$ is strictly convex. Also $K \subset K_2 \subset U$ if $2C_1\epsilon < C_2 - 1$. The proof is therefore complete.
□

A consequence of this result is that any convex, compact set is the decreasing intersection of polyhedra. Also any convex, compact set is the decreasing intersection of smoothly bounded, strongly analytically convex domains.

We know from discussion in Chapter 2 below that any convex domain is the increasing union of smoothly bounded, strongly analytically convex domains. A similar argument shows that any convex domain is the increasing union of polyhedra.

These facts taken together make approximation arguments a very natural vehicle for proving results about convex domains.

The next result is a different sort of approximation argument. It says that any "reasonable" convex set can be written as the union of balls of a uniform size. This type of approximation comes up regularly, for example, in harmonic analysis (see [STE] and also Chapter 8 of [KRA1]).

Let us say that a domain $\Omega$ has $C^{1,1}$ boundary if there is a defining function $\rho$ for $\Omega$ such that $\rho$ is continuously differentiable, and each of the first derivatives of $\rho$ is Lipschitz. See the Appendix. That is to say, there is a constant $C > 0$ such that

$$|\nabla\rho(x) - \nabla\rho(y)| \leq C|x - y|.$$

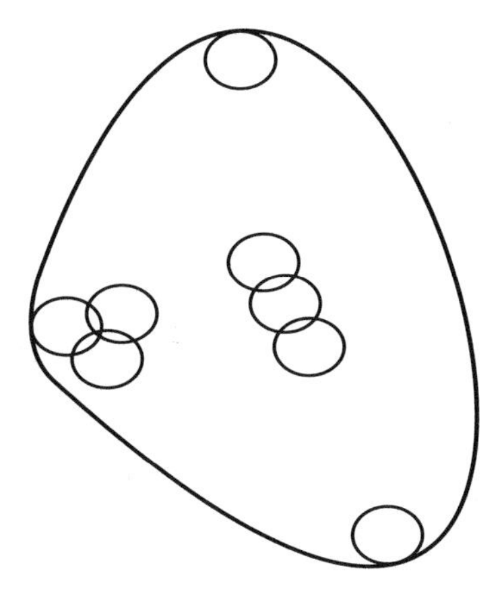

Figure 1.18: Writing a convex set as a union of balls of uniform size.

THEOREM 1.13 *Let $K$ be a compact, convex domain. Then $\partial K$ is $C^{1,1}$ if and only if there is a number $R > 0$ such that $K$ is the union of balls with radius $R$. See Figure 1.18.*

**Proof of the Theorem:** It is useful to write the boundary locally as

$$x_1 = f(x'),$$

with $x' = (x_2, \ldots, x_N)$ and $f$ a $C^{1,1}$ function. Let us assume that $U$ is the domain of $f$, $0 \in U$, $f(0) = 0$, $f'(0) = 0$, and

$$|f'(x) - f'(y)| \leq C|x - y|$$

for all $x, y \in U$. Then certainly

$$|f'(x)| \leq C|x|, \tag{1.13.1}$$

and Taylor's formula then implies that

$$|f(x)| \leq C|x|^2/2. \tag{1.13.2}$$

We may think of the ball with center $(R, 0)$ and radius $R$ as defined by the inequality

$$C|x|^2 \leq 2x_1 R.$$

This implies that, if $C|x'|^2/2 \leq |x|^2/(2R)$, then $f(x') \leq x_1$. This last is true if $CR < 1$. This gives us the ball we seek that contains the origin.

To handle points $p = (f(x'), x')$ near the origin, note that (1.13.1) and (1.13.2) tell us that there is a coordinate change differing from the identity by $(\mathcal{O}(|x'|)$ which shifts the origin to $p$ and the direction $\langle 1, -f'(x')\rangle$ to the direction of the new $x_1$ axis; hence we get the same conclusion (construction of the ball with radius $R$) at this point with $C$ replaced by $C(1+\mathcal{O}(|x'|)) \leq 2C$ if $|x'|$ is small enough.

Let $q$ be any point of $\partial K$. If $R$ is sufficiently small and positive, then there is a ball of radius $R$ contained in $K$ so that $q$ is in the boundary of the ball. Any point $x \in K$ can be written as $x = (1-t)x_1 + tx_2$ with $0 \leq t \leq 1$ and $x_1, x_2 \in \partial K$ (this is just the convexity). If $B_j$ is a ball in $K$ with radius $R$ and having $x_j$ in its boundary, then $(1-t)B_1 + tB_2$ is a ball in $K$ having $x$ in its boundary. That is the ball that we seek.

We leave the details of the converse direction to the reader, or see [HOR, p. 97]. □

## Exercises

1. Prove that the intersection of convex sets is convex.

2. Prove that the set-theoretic product of convex sets is convex.

3. Prove that a convex set is connected, path connected, and simply connected.

4. Is the projection $\pi : \mathbb{R}^3 \to \mathbb{R}^2$, given by $\pi(x_1, x_2, x_3) = (x_1, x_2)$, of a convex set convex?

5. Is the image of a convex set under an invertible linear mapping convex?

6. Prove that the set-theoretic difference $U \setminus V$ of convex sets need not be convex.

7. If $A$ and $B$ are convex sets, then their Minkowski sum is
$$A + B = \{a + b : a \in A, b \in B\}.$$
Prove that the Minkowski sum is convex. [We treat the Minkowski sum in detail in Section 4.2.]

8. A $k$-simplex is a $k$-dimensional polytope which is the convex hull of its k + 1 vertices. Describe all the 2-simplices in the plane. Describe all the 3-simplices in $\mathbb{R}^3$.

9. Refer to Exercise 8 for terminology. Is it true that the increasing union of simplices, if it converges, is still a simplex?

10. What is the difference, in $\mathbb{R}^2$, between a simplex and a polygon with interior?

# Chapter 2

# Characterization of Convexity Using Functions

**Prologue:** Now we begin our analytic studies. The chapter begins with a concept that will be new for many readers: the idea of defining function. This is a key epistemological point in the book—to associate to any reasonable domain a function. The idea is that the function contains all the geometric information about that domain. There are no obvious algebraic operations on domains, but there are many such operations on functions. We can take good advantage of that observation.

Using the defining function, we can finally give an analytic definition of convex set. Advantages now are **(i)** that we can distinguish analytically convex boundary points from non-convex boundary points and **(ii)** that we have a notion of weak analytic convexity and a notion of strong analytic convexity. These are all new ideas, with no precedent in the classical theory of Chapter 1.

We finish the chapter with another new idea—that of exhaustion function. This is a way to characterize a convex set with a function that is defined on the interior of the set only—not on the boundary. From the point of view of intrinsic geometry this is a very natural idea. This discussion rounds out our picture of the study of convex sets using convex functions, and prepares us for deeper explorations in the next chapter.

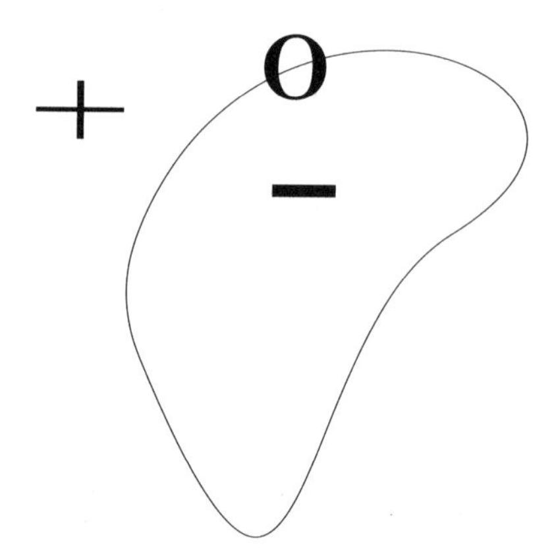


Figure 2.1: The concept of defining function.

## 2.1 The Concept of Defining Function

Let $\Omega \subseteq \mathbb{R}^N$ be a domain with continuously or $C^1$ differentiable boundary. A continuously differentiable function $\rho : \mathbb{R}^N \to \mathbb{R}$ is called a *defining function* for $\Omega$ if

**1.** $\Omega = \{x \in \mathbb{R}^N : \rho(x) < 0\}$;

**2.** ${}^c\overline{\Omega} = \{x \in \mathbb{R}^N : \rho(x) > 0\}$;

**3.** $\nabla\rho(x) \neq 0 \ \ \forall x \in \partial\Omega$.

We note that conditions **1** and **2** certainly entail that $\partial\Omega = \{x \in \mathbb{R}^N : \rho(x) = 0\}$. See Figure 2.1.

In case $k \geq 2$ and $\rho$ is $C^k$ (that is, $k$ times continuously differentiable), then we say that the domain $\Omega$ has $C^k$ boundary. This last point merits some discussion. For the notion of a domain having $C^k$ boundary has many different formulations. One may say that $\Omega$ has $C^k$ boundary if $\partial\Omega$ is a regularly imbedded $C^k$ surface in $\mathbb{R}^N$. Or if $\partial\Omega$ is locally the graph of a $C^k$ function. In the very classical setting of $\mathbb{R}^2$, it is common to say that the boundary of a simply connected domain or region (which of course is just a *curve* $\gamma : S^1 \to \mathbb{R}^2$) is $C^k$ if **(a)** $\gamma$ is a $C^k$ function and **(b)** $\gamma' \neq 0$.

We shall not take the time here to prove the equivalence of all the different formulations of $C^k$ boundary for a domain (but see the rather thorough discussion in Appendix I of [KRA1]). But we do discuss the equivalence of the "local graph" definition with the defining function definition.

First suppose that $\Omega$ is a domain with $C^k$ defining function $\rho$ as specified above, and let $P \in \partial\Omega$. Since $\nabla\rho(P) \neq 0$, the implicit function theorem (see

[KRP2]) guarantees that there is a neighborhood $V_P$ of $P$, a variable (which we may take to be $x_N$) and a $C^k$ function $\varphi_P$ defined on a small open set $U_P \subseteq \mathbb{R}^{N-1}$ so that

$$\begin{aligned}\partial\Omega \cap V_P &= \{(x_1, x_2, \ldots, x_N) : x_N \\ &= \varphi_P(x_1, \ldots, x_{N-1}) \, , \ (x_1, \ldots, x_{N-1}) \in U_P\} \, .\end{aligned}$$

Thus $\partial\Omega$ is locally the graph of the function $\varphi_P$ near $P$.

Conversely, assume that each point $P \in \partial\Omega$ has a neighborhood $V_P$ and an associated $U_P \subseteq \mathbb{R}^{N-1}$ and function $\varphi_P$ such that

$$\begin{aligned}\partial\Omega \cap V_P &= \{(x_1, x_2, \ldots, x_N) : x_N \\ &= \varphi_P(x_1, \ldots, x_{N-1}) \, , \ (x_1, \ldots, x_{N-1}) \in U_P\} \, .\end{aligned}$$

We may suppose that the positive $x_N$-axis points *out* of the domain, and set

$$\rho_P(x) = x_N - \varphi_P(x_1, \ldots, x_{N-1}) \, .$$

Thus, on a small neighborhood of $P$, $\rho_P$ behaves like a defining function. It is equal to 0 on the boundary, certainly has non-vanishing gradient, and is $C^k$.

Now $\partial\Omega$ is compact, so we may cover $\partial\Omega$ with finitely many $V_{P_1}, \ldots, V_{P_k}$. Let $\{\psi_j\}$ be a partition of unity subordinate to this finite cover, and set

$$\widetilde{\rho}(x) = \sum_{j=1}^{k} \psi_j(x) \cdot \rho_{P_j}(x) \, .$$

Then, in a neighborhood of $\partial\Omega$, $\widetilde{\rho}$ is a defining function. We may extend $\widetilde{\rho}$ to all of space as follows. Let $V$ be a neighborhood of $\partial\Omega$ on which $\widetilde{\rho}$ is defined. Let $V'$ be an open, relatively compact subset of $\Omega$ and $V''$ an open subset of ${}^c\overline{\Omega}$ so that $V, V', V''$ cover $\mathbb{R}^N$. Let $\eta, \eta', \eta''$ be a partition of unity subordinate to the cover $V, V', V''$. Now set

$$\begin{aligned}\rho(x) &= \eta'(x) \cdot [-10] \\ &\quad + \eta(x) \cdot \widetilde{\rho}(x) + \eta''(x) \cdot 10 \, .\end{aligned}$$

Then $\rho$ is a globally defined $C^k$ function that is a defining function for $\Omega$.

**Example 2.1** Consider the domain

$$\Omega = \{(x_1, x_2) \in \mathbb{R}^2 : |x_1|^2 + |x_2|^2 < 1\} \, .$$

See Figure 2.2. Obviously this is the unit disc in $\mathbb{R}^2$. It has defining function

$$\rho(x_1, x_2) = |x_1|^2 + |x_1|^2 - 1 \, .$$

We see that $\nabla\rho = \langle 2x_1, 2x_2 \rangle$ is nonvanishing on $\partial\Omega$. And obviously $\rho$ is infinitely differentiable (in fact it is real analytic—that is, it is given by a convergent power series). So $\rho$ is a defining function for $\Omega$ the disc and the disc has $C^\infty$ (indeed real analytic) boundary. ∎

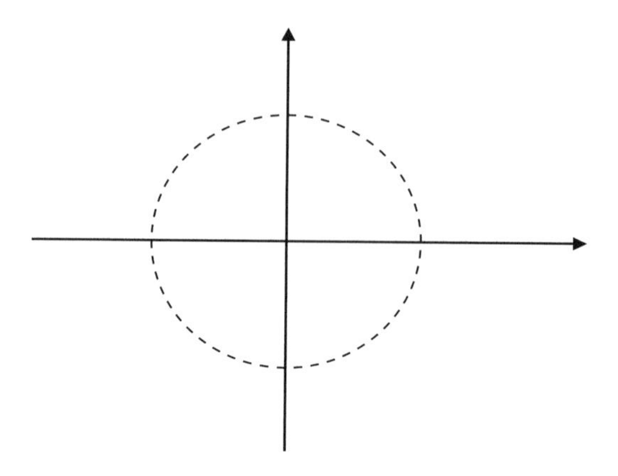

Figure 2.2: The unit disc.

**Example 2.2** Let

$$\Omega = \{(x_1, x_2) \in \mathbb{R}^2 : |x_1| < 1, |x_2| < 1\}.$$

Clearly this is an open unit square in the plane. In particular, it is a domain.

A defining function for this $\Omega$ is

$$\rho(x_1, x_2) = \max\left\{|x_1|, |x_2|\right\} - 1.$$

Notice that $\rho$ is *not* smooth—it is only Lipschitz. But plainly the domain $\Omega$ does not have smooth boundary either. See Figure 2.3. ∎

**Example 2.3** Let $\Omega \subseteq \mathbb{R}^N$ be a bounded domain with $C^k$ boundary, $k \geq 2$, according to some reasonable definition of $C^k$ boundary as discussed above. Let dist denote Euclidean distance. Let $U$ be a tubular neighborhood[1] of $\partial\Omega$. So each point of $U$ has unique nearest point in $\partial\Omega$ (see [HIR] for this idea). Define

$$\psi(x) = \begin{cases} \operatorname{dist}(x, \partial\Omega) & \text{for} \quad x \in {}^c\overline{\Omega} \cap U \\ 0 & \text{for} \quad x \in \partial\Omega \\ -\operatorname{dist}(x, \partial\Omega) & \text{for} \quad x \in \Omega \cap U \end{cases}$$

for $x \in U$. Then $\psi$ behaves like a defining function on $U$. Clearly $\nabla\psi \neq 0$ on $\partial\Omega$ because, for example, the outward normal derivative of $\psi$ on the boundary is 1.

[1]A tubular neighborhood is an open neighborhood of $\partial\Omega$ that is geometrically like a product. See the Appendix.

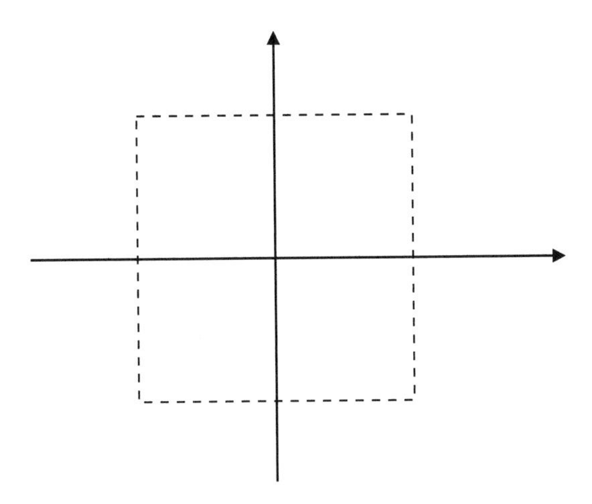

Figure 2.3: The unit square in the plane.

We may extend $\psi$ to all of space in the following way. Let $\varphi$ be a $C_c^\infty$ function supported in $U$ which is identically 1 in a small neighborhood of $\partial\Omega$ and takes values between 0 and 1 inclusive. Define

$$\rho(x) = \begin{cases} \varphi(x)\psi(x) & \text{for} \quad x \in U \\ (1-\varphi(x)) \cdot (-1) & \text{for} \quad x \in \Omega \setminus U \\ (1-\varphi(x)) \cdot (+1) & \text{for} \quad x \in {}^c\Omega \setminus U\,. \end{cases}$$

Then $\rho$ is negative in $\Omega$, positive outside $\overline{\Omega}$, and 0 on $\partial\Omega$. And it is still the case that $\nabla\rho \neq 0$ on $\partial\Omega$. It is a nontrivial theorem that $\psi$, and hence $\rho$, is $C^k$ when the original boundary or $\Omega$ is $C^k$ for $k \geq 2$ (see [KRP3]). Thus we have constructed, by hand, a defining function for $\Omega$. See Figure 2.4. ∎

**Definition 2.4** Let $\Omega \subseteq \mathbb{R}^N$ have $C^1$ boundary and let $\rho$ be a $C^1$ defining function. Let $P \in \partial\Omega$. An $N-$tuple $w = (w_1, \ldots, w_N)$ of real numbers is called a *tangent vector* to $\partial\Omega$ at $P$ if

$$\sum_{j=1}^{N} (\partial\rho/\partial x_j)(P) \cdot w_j = 0.$$

We write $w \in T_P(\partial\Omega)$. See Figure 2.5.

For $\Omega$ with $C^1$ boundary, having defining function $\rho$, we think of

$$\nu_P = \nu = \langle \partial\rho/\partial x_1(P), \ldots, \partial\rho/\partial x_N(P) \rangle$$

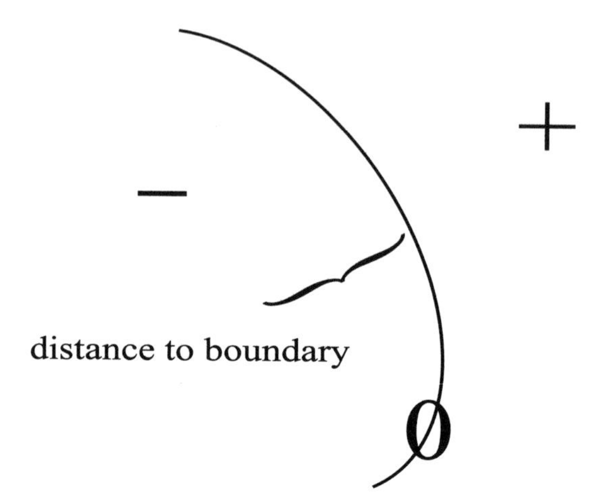


Figure 2.4: The defining function based on distance.

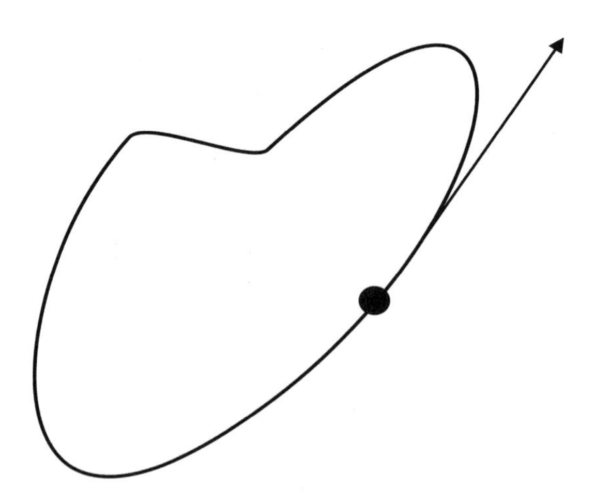

Figure 2.5: A tangent vector.

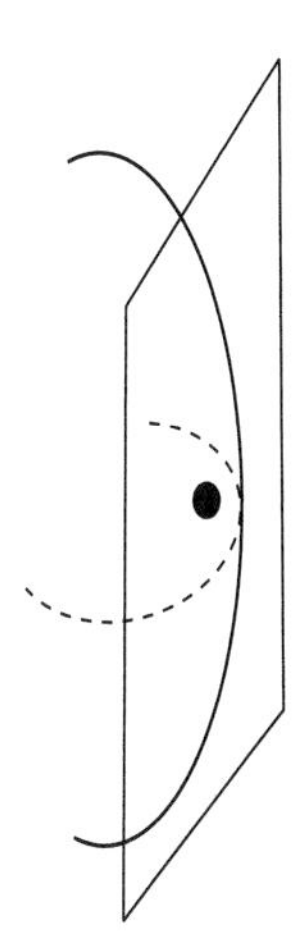

Figure 2.6: The tangent hyperplane.

as the outward-pointing normal vector to $\partial\Omega$ at $P$. Of course the union of all the tangent vectors to $\partial\Omega$ at a point $P \in \partial\Omega$ is the *tangent plane* or *tangent hyperplane* $T_P(\partial\Omega)$. See Figure 2.6. The tangent hyperplane is defined by the condition

$$\nu_P \cdot (w - P) = 0\,.$$

This definition makes sense when $\nu_P$ is well defined, in particular when $\partial\Omega$ is $C^1$.

**Example 2.5** Refer to Example 2.1.

We may calculate that $\partial\rho/\partial x_j = 2x_j$ for $j = 1, 2$. So a vector $w = \langle w_1, w_2\rangle$ is a tangent vector at $P = (p_1, p_2) \in \partial\Omega$ if and only if

$$2p_1w_1 + 2p_2w_2 = 0\,.$$

In other words, if and only if

$$\frac{w_2}{w_1} = -\frac{p_1}{p_2}\,.$$

This makes good geometric sense—see Figure 2.7. ∎

**Example 2.6** Now let

$$\Omega = \{(x_1, x_2, x_3) \in \mathbb{R}^3 : |x_1|^2 + |x_2|^2 + |x_3|^2 < 1\}\,.$$

This is the unit ball in Euclidean 3-space. For specificity, consider the point $P = (1, 0, 0) \in \partial\Omega$. Certainly $\rho(x_1, x_2, x_3) = |x_1|^2 + |x_2|^2 + |x_3|^2 - 1$ is a defining function for $\Omega$. And we see that $\partial\rho/\partial x_j = 2x_j$ for $j = 1, 2, 3$.

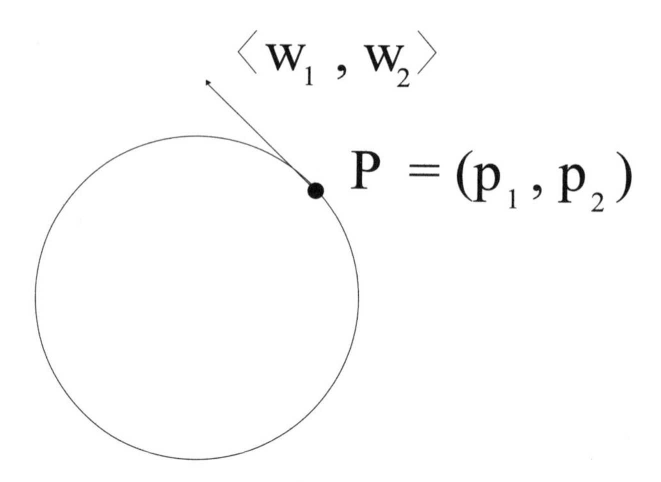


Figure 2.7: A tangent to the boundary of the unit disc in the plane.

We conclude (Figure 2.8) that the vector $w = \langle w_1, w_2, w_3 \rangle$ lies in the tangent space to $\partial\Omega$ at $P$ if and only if

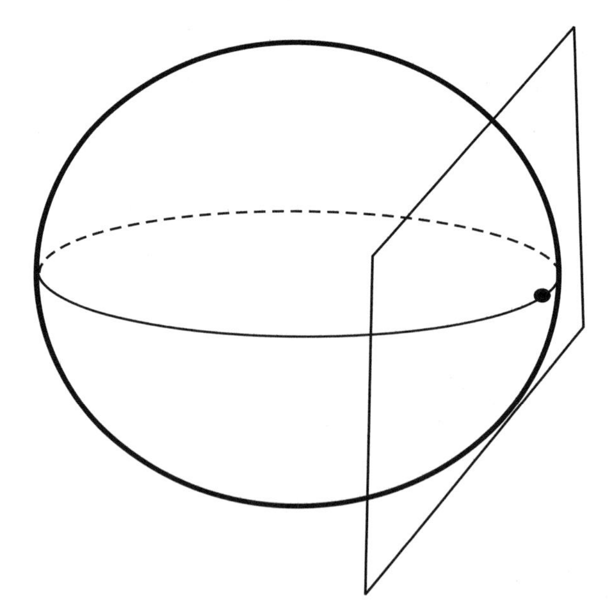

Figure 2.8: The tangent hyperplane to the boundary of the unit ball in $\mathbb{R}^3$ at the point $P$.

$$2 \cdot 1 \cdot (w_1 - 1) + 2 \cdot 0 \cdot w_2 + 2 \cdot 0 \cdot w_3 = 0$$

or $w_1 = 1$. This describes a hyperplane in $\mathbb{R}^3$. ∎

If $\Omega$ is convex and $\partial\Omega$ is not smooth—say that it is Lipschitz—then any point $P \in \partial\Omega$ will still have one (or many) hyperplanes $\mathcal{P}$ such that $\mathcal{P} \cap \overline{\Omega} = \{P\}$. We call such a hyperplane a *support hyperplane* for $\partial\Omega$ at $P$. As noted, such a support hyperplane need not be unique. For example, if $\Omega = \{(x_1, x_2) : |x_1| < 1, |x_2| < 1\}$ then the points of the form $(\pm 1, \pm 1)$ in the boundary do *not* have well-defined tangent lines, but they do have (uncountably) many support lines.

Of course the definition of the normal $\nu_P$ makes sense only if it is independent of the choice of $\rho$. We shall address that issue in a moment. It should be observed that the condition defining tangent vectors $w$ simply mandates that $w \perp \nu_P$ at $P$. And, after all, we know from calculus that $\nabla\rho$ is the normal $\nu_P$ and that the normal is uniquely determined and independent of the choice of $\rho$. In principle, this settles the well-definedness issue.

However this point is so important, and the point of view that we are considering so pervasive, that further discussion is warranted. The issue is this: if $\widehat{\rho}$ is another defining function for $\Omega$ then it should give the same tangent vectors as $\rho$ at any point $P \in \partial\Omega$. The key to seeing that this is so is to write $\widehat{\rho}(x) = h(x) \cdot \rho(x)$, for $h$ a smooth function that is non-vanishing near $\partial\Omega$. Then, for $P \in \partial\Omega$,

$$\begin{aligned}
\sum_{j=1}^{N} (\partial\widehat{\rho}/\partial x_j)(P) \cdot w_j \\
&= h(P) \cdot \left( \sum_{j=1}^{N} (\partial\rho/\partial x_j)(P) \cdot w_j \right) \\
&\quad + \rho(P) \cdot \left( \sum_{j=1}^{N} (\partial h/\partial x_j)(P) \cdot w_j \right) \\
&= h(P) \cdot \left( \sum_{j=1}^{N} (\partial\rho/\partial x_j)(P) \cdot w_j \right) \\
&\quad + 0,
\end{aligned} \tag{2.7}$$

because $\rho(P) = 0$. Thus $w$ is a tangent vector at $P$ vis à vis $\rho$ if and only if $w$ is a tangent vector vis à vis $\widehat{\rho}$. But why does $h$ exist?

After a change of coordinates, it is enough to assume that we are dealing with a piece of $\partial\Omega$ that is a piece of flat, $(N-1)$–dimensional real hypersurface (just use the implicit function theorem). Thus we may take $\rho(x) = x_N$ and $P = 0$. Then any other defining function $\widehat{\rho}$ for $\partial\Omega$ near $P$ must have the Taylor expansion

$$\widehat{\rho}(x) = c \cdot x_N + \mathcal{R}(x) \tag{2.8}$$

about 0. Here $\mathcal{R}$ is a remainder term[2] satisfying $\mathcal{R}(x) = o(|x_N|)$. There is no loss of generality to take $c = 1$, and we do so in what follows. Thus we wish to define

$$h(x) = \frac{\widehat{\rho}(x)}{\rho(x)} = 1 + \mathcal{S}(x).$$

Here $\mathcal{S}(x) \equiv \mathcal{R}(x)/x_N$ and $\mathcal{S}(x) = o(1)$ as $x_N \to 0$. Since this remainder term involves a derivative of $\widehat{\rho}$, it is plain that $h$ is not even differentiable. (An explicit counterexample is given by $\widehat{\rho}(x) = x_N \cdot (1 + |x_N|)$.) Thus the program that we attempted in equation (2.7) above is apparently flawed.

However an inspection of the explicit form of the remainder term $\mathcal{R}$ reveals that, because $\widehat{\rho}$ is constant on $\partial\Omega$, $h$ as defined above *is* continuously differentiable *in tangential directions.* That is, for tangent vectors $w$ (vectors that are orthogonal to $\nu_P$), the derivative

$$\sum_j \frac{\partial h}{\partial x_j}(P) w_j$$

*is* defined. Thus it does indeed turn out that our definition of tangent vector is well posed when it is applied to vectors *that are already known to be tangent vectors* by the geometric definition $w \cdot \nu_P = 0$. For vectors that are *not* geometric tangent vectors, an even simpler argument shows that

$$\sum_j \frac{\partial \widehat{\rho}}{\partial x_j}(P) w_j \neq 0$$

if and only if

$$\sum_j \frac{\partial \rho}{\partial x_j}(P) w_j \neq 0.$$

Thus Definition 2.4 is well posed. Questions similar to the one just discussed will come up below when we define convexity using $C^2$ defining functions. They are resolved in just the same way and we shall leave details to the reader.

The reader should check that the discussion above proves the following: if $\rho, \widetilde{\rho}$ are $C^k$ defining functions for a domain $\Omega$, with $k \geq 2$, then there is a

[2]We may think of (2.8) as proved by integration by parts in the $x_N$ variable only, and that gives this favorable estimate on the error terms $\mathcal{R}(x)$.

$C^{k-1}$, nonvanishing function $h$ defined near $\partial\Omega$ such that $\rho = h \cdot \widetilde{\rho}$.

## 2.2 The Analytic Definition of Convexity

For convenience, we restrict attention for this section to *bounded* domains. Many of our definitions would need to be modified, and extra arguments given in proofs, were we to consider unbounded domains as well.

**Definition 2.9** Let $\Omega \subseteq \mathbb{R}^N$ be a bounded domain with $C^2$ boundary and $\rho$ a defining function for $\Omega$. Fix a point $P \in \partial\Omega$. We say that $\partial\Omega$ is analytically (weakly) *convex* at $P$ if

$$\sum_{j,k=1}^{N} \frac{\partial^2 \rho}{\partial x_j \partial x_k}(P) w_j w_k \geq 0, \quad \forall w \in T_P(\partial\Omega).$$

We say that $\partial\Omega$ is analytically *strongly (strictly) analytically convex* at $P$ if the inequality is strict whenever $w \neq 0$.

If $\partial\Omega$ is analytically convex (resp. strongly analytically convex) at each boundary point then we say that $\Omega$ is analytically convex (resp. strongly analytically convex).

One interesting and useful feature of this new definition of convexity is that it treats the concept point by point. The classical, synthetic (geometric) definition specifies convexity for the whole domain at once. See Figure 2.9.

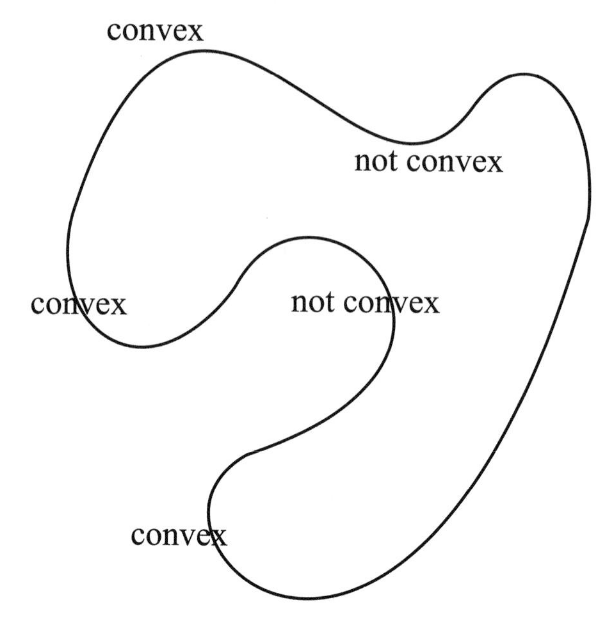


Figure 2.9: Convexity defined point by point.

**Example 2.10** Consider the domain

$$\Omega = \{x \in \mathbb{R}^2 : x_1^2 + x_2^2 < 1\} .$$

This domain (the unit disc) is strongly analytically convex. By contrast, the domain

$$\widehat{\Omega} = \{x \in \mathbb{R}^2 : x_1^2 + x_2^4 < 1\}$$

is weakly (but not strongly) analytically convex. The weak convexity occurs at the boundary points $(\pm 1, 0)$. In fact you can calculate the matrix of second partials for $\widehat{\Omega}$ at $(1, 0)$ to be

$$\begin{pmatrix} 1 & 0 \\ 0 & 0 \end{pmatrix}$$

The tangent vector at $(1, 0)$ is $\langle 0, 1 \rangle$. And thus our expression that defines convexity at $(1, 0)$ is

$$1 \cdot 0 \cdot 0 + 0 \cdot 0 \cdot 1 + 0 \cdot 1 \cdot 0 + 0 \cdot 1 \cdot 1 = 0 .$$

So it is plain that the boundary point $(1, 0)$ is weakly (not strongly) analytically convex.

∎

It is natural to ask whether the new definition of convexity is independent of the choice of defining function. We have the following result:

PROPOSITION 2.11 *Let $\Omega \subseteq \mathbb{R}^N$ be a domain with $C^2$ boundary. Let $\rho$ and $\widetilde{\rho}$ be $C^2$ defining functions for $\Omega$, and assume that, at points $x$ near $\partial\Omega$,*

$$\rho(x) = h(x) \cdot \widetilde{\rho}(x)$$

*for some non-vanishing, $C^2$ function $h$. Let $P \in \partial\Omega$. Then $\Omega$ is analytically convex at $P$ when measured with the defining function $\rho$ if and only if $\Omega$ is analytically convex at $P$ when measured with the defining function $\widetilde{\rho}$.*

**Proof:** We calculate that

$$\begin{aligned}
\frac{\partial^2}{\partial x_j \partial x_k}\rho(P) &= \\
& h(P) \cdot \frac{\partial^2 \widetilde{\rho}}{\partial x_j \partial x_k}(P) \\
& + \widetilde{\rho}(P) \cdot \frac{\partial^2 h}{\partial x_j \partial x_k}(P) \\
& + \frac{\partial \widetilde{\rho}}{\partial x_j}(P)\frac{\partial h}{\partial x_k}(P) \\
& + \frac{\partial \widetilde{\rho}}{\partial x_k}(P)\frac{\partial h}{\partial x_j}(P) \\
&= h(P) \cdot \frac{\partial^2 \widetilde{\rho}}{\partial x_j \partial x_k}(P) \\
& + \frac{\partial \widetilde{\rho}}{\partial x_j}(P)\frac{\partial h}{\partial x_k}(P) \\
& + \frac{\partial \widetilde{\rho}}{\partial x_k}(P)\frac{\partial h}{\partial x_j}(P)
\end{aligned}$$

because $\widetilde{\rho}(P) = 0$. But then, if $w$ is a tangent vector to $\partial\Omega$ at $P$, we see that

$$\begin{aligned}
& \sum_{j,k} \frac{\partial^2}{\partial x_j \partial x_k}\rho(P) w_j w_k \\
&= h(P) \sum_{j,k} \frac{\partial^2 \widetilde{\rho}}{\partial x_j \partial x_k}(P) w_j w_k \\
&\quad + \left[\sum_j \frac{\partial \widetilde{\rho}}{\partial x_j}(P) w_j\right] \left[\sum_k \frac{\partial h}{\partial x_k}(P) w_k\right] \\
&\quad + \left[\sum_k \frac{\partial \widetilde{\rho}}{\partial x_k}(P) w_k\right] \left[\sum_j \frac{\partial h}{\partial x_j}(P) w_j\right] .
\end{aligned}$$

If we suppose that $P$ is a point of analytic convexity relative to the defining function $\widetilde{\rho}$, then the first sum is nonnegative. Of course $h$ is positive, so the first expression is then $\geq 0$. Since $w$ is a tangent vector, the sum in $j$ in the second expression vanishes. Likewise the sum in $k$ in the third expression vanishes.

In the end, we see that the Hessian of $\rho$ is positive semi-definite on the tangent space if the Hessian of $\widetilde{\rho}$ is. The reasoning also works if the roles of $\rho$ and $\widetilde{\rho}$ are reversed. The result is thus proved. □

The quadratic form

$$\left(\frac{\partial^2 \rho}{\partial x_j \partial x_k}(P)\right)_{j,k=1}^{N}$$

is frequently called the "real Hessian" of the function $\rho$. This form carries considerable geometric information about the boundary of $\Omega$. It is of course closely related to the second fundamental form of Riemannian geometry (see B. O'Neill [ONE]).

It is common to use either of the words "strong" or "strict" to mean that the inequality in the last definition is strict when $w \neq 0$. The reader may wish to verify that, at a strongly analytically convex boundary point, all curvatures are positive (in fact one may, by the positive definiteness of the matrix $\left(\partial^2 \rho / \partial x_j \partial x_k\right)$, impose a change of coordinates at $P$ so that the boundary of $\Omega$ agrees with a ball up to second order at $P$).

Now we explore our analytic notions of convexity. The first lemma is a technical one:

LEMMA 2.12 *Let $\Omega \subseteq \mathbb{R}^N$ be strongly analytically convex. Then there is a constant $C > 0$ and a defining function $\widetilde{\rho}$ for $\Omega$ such that*

$$\sum_{j,k=1}^{N} \frac{\partial^2 \widetilde{\rho}}{\partial x_j \partial x_k}(P) w_j w_k \geq C|w|^2, \quad \forall P \in \partial\Omega, w \in \mathbb{R}^N. \tag{2.12.1}$$

We note that the lemma says the following:

> At a strongly convex point, we can choose a defining function so that the quadratic form induced by the Hessian is positive definite *on all vectors*—not just tangent vectors.

This is a remarkable result—and a useful one. For it enables us to see right away that strong analytic convexity is a stable property, that is to say, stable under perturbation. Put in other words, if a boundary point $P$ is strongly analytically convex then nearby points will also be strongly analytically convex.

**Proof of the Lemma:** Let $\rho$ be some fixed $C^2$ defining function for $\Omega$. For $\lambda > 0$ define

$$\rho_\lambda(x) = \frac{\exp(\lambda\rho(x)) - 1}{\lambda}.$$

We shall select the constant $\lambda$ large in a moment. Let $P \in \partial\Omega$ and set

$$X = X_P = \left\{ w \in \mathbb{R}^N : |w| = 1 \text{ and } \sum_{j,k} \frac{\partial^2 \rho}{\partial x_j \partial x_k}(P) w_j w_k \leq 0 \right\}.$$

Then no element of $X$ could be a tangent vector at $P$, hence $X \subseteq \{w : |w| = 1 \text{ and } \sum_j \partial\rho/\partial x_j(P) w_j \neq 0\}$. Since $X$ is defined by a non-strict inequality, it is closed; it is of course also bounded. Hence $X$ is compact and

$$\mu \equiv \min \left\{ \left| \sum_j \partial\rho/\partial x_j(P) w_j \right| : w \in X \right\}$$

is attained and is non-zero. Define

$$\lambda = \frac{-\min_{w \in X} \sum_{j,k} \frac{\partial^2 \rho}{\partial x_j \partial x_k}(P) w_j w_k}{\mu^2} + 1.$$

Set $\widetilde{\rho} = \rho_\lambda$. Then for any $w \in \mathbb{R}^N$ with $|w| = 1$ we have (since $\exp(\rho_(P)) = 1$) that

$$\begin{aligned}
&\sum_{j,k} \frac{\partial^2 \widetilde{\rho}}{\partial x_j \partial x_k}(P) w_j w_k \\
&= \sum_{j,k} \left\{ \frac{\partial^2 \rho}{\partial x_j \partial x_k}(P) + \lambda \frac{\partial \rho}{\partial x_j}(P) \frac{\partial \rho}{\partial x_k}(P) \right\} w_j w_k \\
&= \sum_{j,k} \left\{ \frac{\partial^2 \rho}{\partial x_j \partial x_k} \right\} (P) w_j w_k + \lambda \left| \sum_j \frac{\partial \rho}{\partial x_j}(P) w_j \right|^2
\end{aligned}$$

If $w \notin X$ then this expression is positive by definition. If $w \in X$ then the expression is positive by the choice of $\lambda$. Since $\{w \in \mathbb{R}^N : |w| = 1\}$ is compact, there is thus a $C > 0$ such that

$$\sum_{j,k} \left\{ \frac{\partial^2 \widetilde{\rho}}{\partial x_j \partial x_k} \right\} (P) w_j w_k \geq C,$$

$$\forall w \in \mathbb{R}^N \text{ such that } |w| = 1.$$

This establishes our inequality (2.12.1) for $P \in \partial\Omega$ fixed and $w$ in the unit sphere of $\mathbb{R}^N$. For arbitrary $w$, we set $w = |w|(w/|w|) = |w|\widehat{w}$, with $\widehat{w}$ in the unit sphere. Then (2.12.1) holds for $\widehat{w}$:

$$\sum_{j,k=1}^{N} \frac{\partial^2 \widetilde{\rho}}{\partial x_j \partial x_k}(P)\widehat{w}_j \widehat{w}_k \geq C|\widehat{w}|^2 .$$

Multiplying both sides of the inequality by $|w|^2$ now gives

$$\sum_{j,k=1}^{N} \frac{\partial^2 \widetilde{\rho}}{\partial x_j \partial x_k}(P) w_j w_k \geq C|w|^2$$

and that is what we wished to prove.

Finally, notice that our estimates, in particular the existence of $C$, hold uniformly over points in $\partial\Omega$ near $P$. Since $\partial\Omega$ is compact, we see that the constant $C$ may be chosen uniformly over all boundary points of $\Omega$. □

Notice that the statement of the lemma has two important features: **(i)** that the constant $C$ may be selected uniformly over the boundary and **(ii)** that the inequality (2.12.1) holds for all $w \in \mathbb{R}^N$ (not just tangent vectors). In fact it is impossible to arrange for anything like (2.12.1) to be true at a weakly analytically convex point.

Our proof shows in fact that (2.12.1) is true not just for $P \in \partial\Omega$ but for $P$ in a neighborhood of $\partial\Omega$. It is this sort of stability of the notion of strong analytic convexity that makes it a more useful device than ordinary (weak) analytic convexity.

PROPOSITION 2.13 *If $\Omega$ is strongly analytically convex then $\Omega$ is geometrically convex.*

**Proof:** We use a connectedness argument.

Clearly $\Omega \times \Omega$ is connected. Set $S = \{(P_1, P_2) \in \Omega \times \Omega : (1-\lambda)P_1 + \lambda P_2 \in \Omega,\ \text{all } 0 < \lambda < 1\}$. Then $S$ is plainly open and non-empty.

To see that $S$ is closed, fix a defining function $\widetilde{\rho}$ for $\Omega$ as in the lemma. If $S$ is not closed in $\Omega \times \Omega$ then there exist $P_1, P_2 \in \Omega$ such that the function

$$t \mapsto \widetilde{\rho}((1-t)P_1 + tP_2)$$

assumes an interior maximum value of 0 on $[0,1]$. But the positive definiteness of the real Hessian of $\widetilde{\rho}$ contradicts that assertion. The proof is complete. □

We gave a special proof that strong convexity implies geometric convexity simply to illustrate the utility of the strong convexity concept. It is possible to prove that an arbitrary (weakly) analytically convex domain is geometrically convex by showing that such a domain can be written as the increasing union of strongly analytically convex domains. However the proof is difficult and technical. We thus give another proof of this fact:

PROPOSITION 2.14 *If $\Omega$ is (weakly) analytically convex then $\Omega$ is geometrically convex.*

**Proof:** To simplify the proof we shall assume that $\Omega$ has at least $C^3$ boundary.

Assume without loss of generality that $N \geq 2$ and $0 \in \Omega$. For $\epsilon > 0$, let $\rho_\epsilon(x) = \rho(x) + \epsilon|x|^{2M}/M$ and $\Omega_\epsilon = \{x : \rho_\epsilon(x) < 0\}$. Then $\Omega_\epsilon \subseteq \Omega_{\epsilon'}$ if $\epsilon' < \epsilon$ and $\cup_{\epsilon>0}\Omega_\epsilon = \Omega$. If $M \in \mathbb{N}$ is large and $\epsilon$ is small, then $\Omega_\epsilon$ is strongly analytically convex. By Proposition 2.13, each $\Omega_\epsilon$ is geometrically convex, so $\Omega$ is convex. □

We mention in passing that a nice treatment of convexity, from roughly the point of view presented here, appears in [VLA].

PROPOSITION 2.15 *Let $\Omega \subseteq \mathbb{R}^N$ have $C^2$ boundary and be geometrically convex. Then $\Omega$ is (weakly) analytically convex.*

**Proof:** Seeking a contradiction, we suppose that for some $P \in \partial\Omega$ and some $w \in T_P(\partial\Omega)$ we have

$$\sum_{j,k} \frac{\partial^2 \rho}{\partial x_j \partial x_k}(P) w_j w_k = -2K < 0. \tag{2.15.1}$$

Suppose without loss of generality that coordinates have been selected in $\mathbb{R}^N$ so that $P = 0$ and $(0, 0, \ldots, 0, 1)$ is the unit outward normal vector to $\partial\Omega$ at $P$. We may further normalize the defining function $\rho$ so that $\partial\rho/\partial x_N(0) = 1$. Let $Q = Q^t = tw + \epsilon \cdot (0, 0, \ldots, 0, 1)$, where $\epsilon > 0$ and $t \in \mathbb{R}$. Then, by Taylor's expansion,

$$
\begin{aligned}
\rho(Q) &= \rho(0) + \sum_{j=1}^{N} \frac{\partial \rho}{\partial x_j}(0) Q_j \\
&\quad + \frac{1}{2} \sum_{j,k=1}^{N} \frac{\partial^2 \rho}{\partial x_j \partial x_k}(0) Q_j Q_k + o(|Q|^2) \\
&= \epsilon \frac{\partial \rho}{\partial x_N}(0) + \frac{t^2}{2} \sum_{j,k=1}^{N} \frac{\partial^2 \rho}{\partial x_j \partial x_k}(0) w_j w_k \\
&\quad + \mathcal{O}(\epsilon^2) + o(t^2) \\
&= \epsilon - K t^2 + \mathcal{O}(\epsilon^2) + o(t^2) .
\end{aligned}
$$

Thus, if $t = 0$ and $\epsilon > 0$ is small enough, then $\rho(Q) > 0$. However, for that same value of $\epsilon$, if $|t| > \sqrt{2\epsilon/K}$ then $\rho(Q) < 0$. This contradicts the definition of geometric convexity. □

**Remark 2.16** The reader can already see in the proof of the proposition how useful the quantitative version of convexity can be.

The assumption that $\partial\Omega$ be $C^2$ is not very restrictive, for convex functions of one variable are twice differentiable almost everywhere (see [ZYG] and [EVG]). On the other hand, $C^2$ smoothness of the boundary is essential for our approach to the subject.

**Exercises for the Reader:** Assume in the following problems that $\overline{\Omega} \subseteq \mathbb{R}^N$ is closed, bounded, and convex. Assume that $\Omega$ has $C^2$ boundary.

**(a)** We shall say more about extreme points later in the book. For now, a point $P \in \partial\Omega$ is extreme (for $\Omega$ convex) if, whenever $P = (1 - \lambda)x + \lambda y$ and $0 \leq \lambda \leq 1$, $x, y \in \overline{\Omega}$, then $x = y = P$. See Figures 2.10 and 2.11.

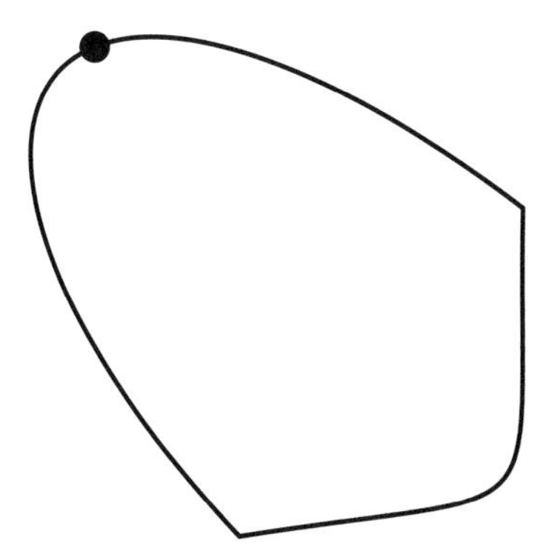

Figure 2.10: An extreme point.

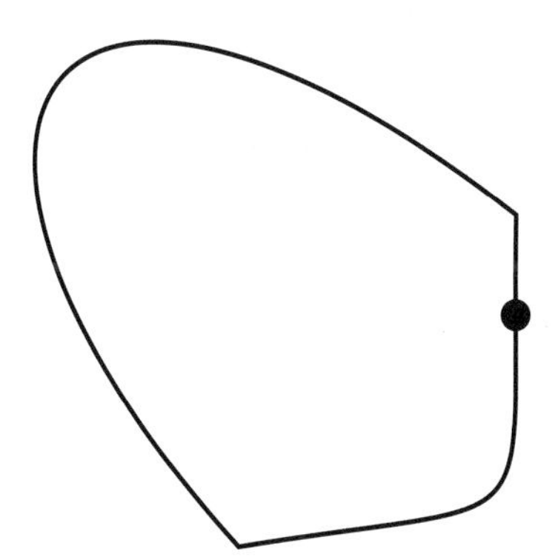

Figure 2.11: A non-extreme point.

Prove that $\overline{\Omega}$ is the closed convex hull of its extreme points (this result is usually referred to as the *Krein-Milman theorem* and is true in much greater generality). We shall treat this remarkable theorem in greater detail below.

(**b**) Let $P \in \partial\Omega$ be extreme. Let $\mathbf{p} = P + T_P(\partial\Omega)$ be the geometric tangent affine hyperplane to the boundary of $\Omega$ that passes through $P$. Show by an example that it is not necessarily the case that $\mathbf{p} \cap \overline{\Omega} = \{P\}$.

(**c**) Prove that, if $\Omega_0$ is *any* bounded domain with $C^2$ boundary, then there is a relatively open subset $U$ of $\partial\Omega_0$ such that $U$ is strongly analytically convex. (Hint: Fix $x_0 \in \Omega_0$ and choose $P \in \partial\Omega_0$ that is as far as possible from $x_0$). See Figure 2.12.

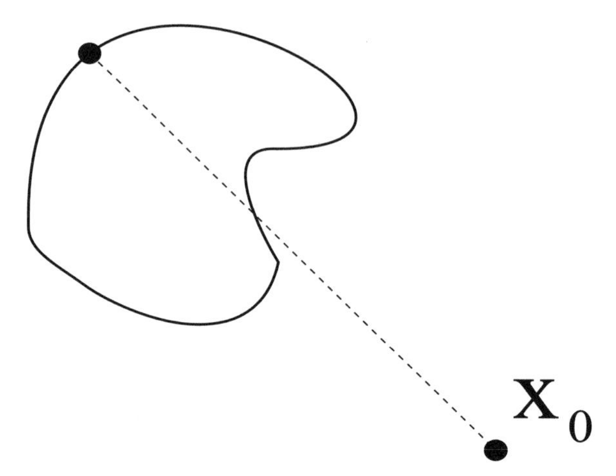


Figure 2.12: Existence of strongly analytically convex boundary points.

**(d)** If $\Omega$ is a convex domain then the Minkowski functional[3] (see [LAY]) less 1 gives a convex defining function for $\Omega$.

The new notions of (weakly) analytically convex and strongly analytically convex give us a quantitative, differential-geometric way to think about convexity. As the book proceeds, we shall see that this is a powerful tool that will enable us to discover new ideas, and new results, about convexity.

## 2.3 Convex Functions

**Definition 2.17** Let $F : \mathbb{R}^N \to \mathbb{R}$ be a function. We say that $F$ is *convex* if, for any $P, Q \in \mathbb{R}^N$ and any $0 \leq t \leq 1$, it holds that

$$F((1-t)P + tQ) \leq (1-t)F(P) + tF(Q) .$$

See Figure 2.13. We say that $F$ is *strictly (or strongly) convex* if the inequality is strict.

**Example 2.18** The function $f(x) = e^x$ is strictly convex.

To see this, consider the inequality

$$f((1-t)P + tQ) \leq (1-t)f(P) + tf(Q) .$$

[3] A simple instance of the Minkowski functional is the following. Let $K \subseteq \mathbb{R}^N$ be convex. For $x \in \mathbb{R}^N$, define

$$p(x) = \inf\{r > 0 : x \in rK\} .$$

Then $p$ is a Minkowski functional for $K$. The Minkowski functional is an instance of a gauge, a concept that is discussed in Section 5.1.

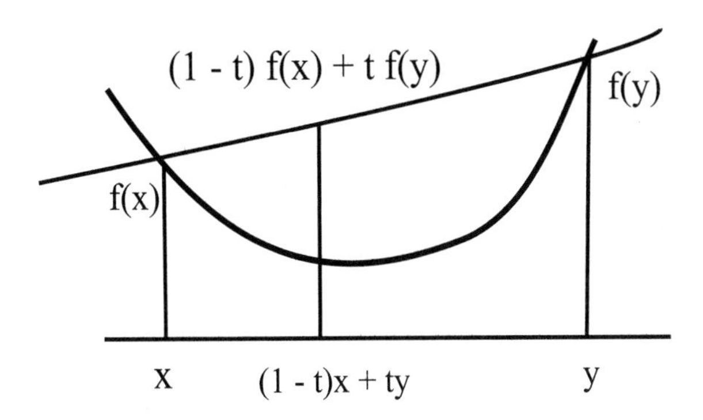


Figure 2.13: The definition of convex function.

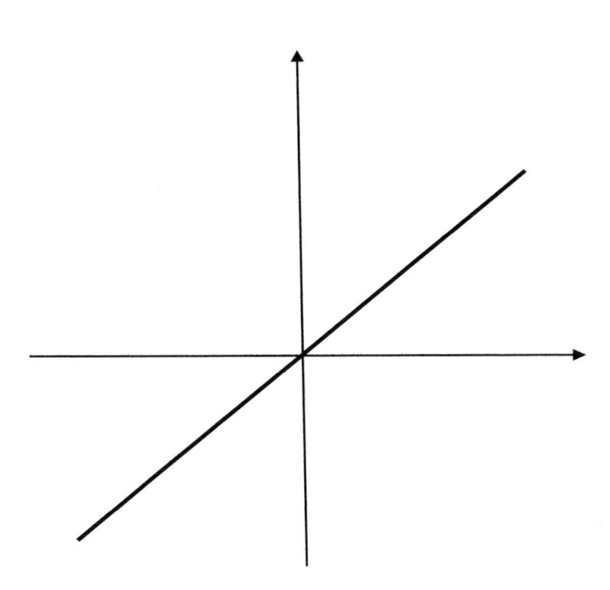

Figure 2.14: A convex function that is not strictly convex.

This becomes

$$e^{(1-t)P+tQ} \le (1-t)e^P + te^Q$$

or

$$e^{t(Q-P)} \le (1-t) + te^{Q-P}.$$

Set $\alpha = Q - P$ and, for $t$ fixed, $A(\alpha) = e^{t\alpha}$, $B(\alpha) = (1-t) + te^{\alpha}$.

Notice that $A(0) = B(0)$ and $A' = te^{t\alpha}$, $B' = te^{\alpha}$. So it is clear that $A' < B'$. We conclude then that $A < B$ and the function is strictly convex. ∎

**Example 2.19** The function $g(x) = x$ is convex but *not* strictly convex. See Figure 2.14. ∎

PROPOSITION 2.20 *A strictly convex function defined on an open interval $I \subseteq \mathbb{R}$ has at most one minimum.*

**Proof:** To see this, suppose not. Let $p < q$ be minima for a strictly convex function $f$. Assume without loss of generality that $f(p) \le f(q)$. Then the inequality

$$f((1-t)P + tQ) < (1-t)f(P) + tf(Q)$$

tells us that, for $x$ near to and to the left of $q$, $f(x) < f(q)$. This contradicts the fact that $q$ is a minimum and completes the proof. □

The same reasoning shows that a strictly convex function on an open interval has no maximum. What can you say about a non-strictly convex (i.e., a weakly convex) function?

A special form of convex function is the following:

**Definition 2.21** We say that $F : \mathbb{R}^n \to \mathbb{R}$ is *midpoint convex* if, for any $P, Q \in \mathbb{R}^N$,

$$F((1/2)P + (1/2)Q) \le (1/2)F(P) + (1/2)F(Q). \tag{2.21.1}$$

Clearly any convex function is midpoint convex. The converse is true for any function that is known to be continuous. We omit the details, but refer the reader to [SIM, p. 3].

**Example 2.22** Consider the function $f(x) = e^x$. Checking line (2.21.1) amounts to

$$e^{P/2+Q/2} \le \frac{1}{2}e^P + \frac{1}{2}e^Q,$$

and this is true because

$$2ab \le a^2 + b^2.$$

Thus $f$ is midpoint convex. Since it is obviously continuous, it is therefore convex. [This repeats the result of Example 2.18.] ∎

**Example 2.23** Suppose that $g : \mathbb{R} \to \mathbb{R}$ is convex and increasing. Then the function

$$f(x) = g(|x|)$$

is convex.

To see this, we calculate that

$$\begin{aligned} f((1-t)x + ty) &= g(|(1-t)x + ty|) \\ &\le g((1-t)|x| + t|y|) \\ &\le (1-t)g(|x|) + tg(|y|) \\ &= (1-t)f(x) + tf(y) \,. \end{aligned}$$

∎

**Example 2.24** Let $p > 0$ and $f(x) = |x|^p$. Is this function convex?

By the preceding example, we need only consider the case $x \ge 0$. By continuity, we may further restrict attention to $x > 0$. So we may simply differentiate.

Thus

$$f''(x) = p(p-1)|x|^{p-2} \,.$$

This is positive when $p > 1$, so we see that in that circumstance $f$ is strictly convex. When $p = 1$, then $f$ is convex but not strictly so. For $p < 1$, $f$ is *not* convex. We encourage the reader to draw some graphs to illustrate these ideas. ∎

PROPOSITION 2.25 *Let $\{f_\alpha\}$ be convex functions on an open, convex set $U \subseteq \mathbb{R}^N$. Define*

$$f(x) = \sup_\alpha f_\alpha(x)$$

*for each $x \in U$. Then $f$ is convex.*

**Proof:** This is a straightforward calculation:

$$\begin{aligned} f_\alpha((1-t)x + ty) &\le (1-t)f_\alpha(x) + tf_\alpha(y) \\ &\le (1-t)f(x) + tf(y) \,. \end{aligned}$$

Now, passing to the supremum in $\alpha$ on the left-hand side, we get our result. □

The next result is geometrically clear (see Figure 2.15), but the proof actually requires some effort.

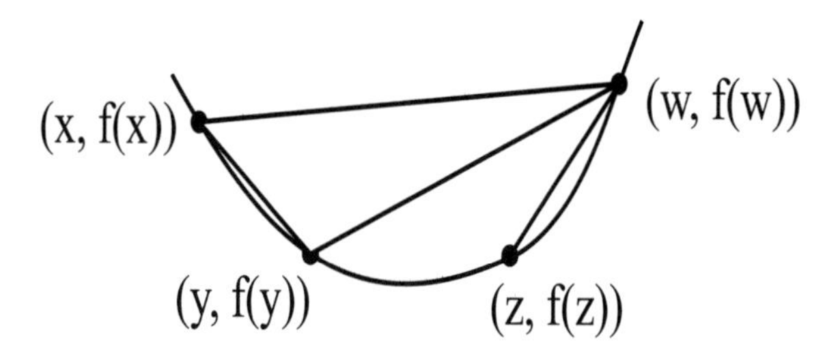


Figure 2.15: Comparing slopes.

PROPOSITION 2.26 *Let $f$ be a convex function on an interval $I \subseteq \mathbb{R}$. Let $x, y, z, w \in I$ with $x < y < z < w$. Then*

$$\frac{f(y) - f(x)}{y - x} \leq \frac{f(w) - f(z)}{w - z}.$$

**Proof:** First suppose that $y = z$. We write $y$ as a convex combination of $x$ and $w$, thus

$$y = \left(\frac{w - y}{w - x}\right) + \left(\frac{y - x}{w - x}\right) w.$$

Therefore

$$\begin{aligned} f(y) &\leq \left(\frac{w - y}{w - x}\right) f(x) + \left(\frac{y - x}{w - x}\right) f(w) \\ &= f(x) + \left(\frac{y - x}{w - x}\right) \cdot [f(w) - f(x)]. \end{aligned} \tag{2.26.1}$$

This implies that

$$\frac{f(y) - f(x)}{y - x} \leq \frac{f(w) - f(x)}{w - x}.$$

We may rewrite (2.26.1) as

$$f(y) \leq f(x) + \left(\frac{y - x}{w - x}\right) [f(w) - f(y) + f(y) - f(x)].$$

Therefore

$$[f(y) - f(x)] \cdot \left(1 - \frac{y - x}{w - x}\right) \leq \left(\frac{y - x}{w - x}\right) [f(w) - f(y)],$$

which is the same as

$$[f(y) - f(x)] \left(\frac{w - y}{w - x}\right) \leq \left(\frac{y - x}{w - x}\right) \cdot [f(w) - f(y)].$$

This implies our result in the special case that $y = z$.

For the general case of the proposition, we apply the special case twice:

$$\frac{f(y) - f(x)}{y - x} \leq \frac{f(z) - f(y)}{z - y} \leq \frac{f(w) - f(z)}{w - z}. \qquad \square$$

In the case that $F$ is $C^2$, we may restrict $F$ to the line passing through $P$ and $Q$ and differentiate the function

$$\varphi_{P,Q} : t \longmapsto F((1-t)P + tQ)$$

twice to see (from calculus—reference [BLK]) that

$$\frac{d^2}{dt^2}\varphi_{P,Q} \geq 0 .$$

If we set $\alpha = Q - P = (\alpha_1, \alpha_2, \ldots, \alpha_N)$, then this last result may be written as

$$\sum_{j,k} \frac{\partial^2}{\partial x_j \partial x_k} \alpha_j \alpha_k \geq 0 .$$

In other words, the Hessian of $F$ is positive semi-definite.

It is *not* the case that positive definite Hessian corresponds to strict convexity, as the following example shows.

**Example 2.27** Consider the function $f(x) = x^4$. Then $f''(x) = 12x^2$, and obviously $f''(0) = 0$. Nevertheless, $f$ is strictly convex.

To see this, we need to check that

$$\left[(1-t)x + ty\right]^4 < (1-t)x^4 + ty^4 .$$

We rewrite this as

$$\left[(1-t) + t\frac{y}{x}\right] < (1-t) + t\left(\frac{y}{x}\right)^4 .$$

Now set $\alpha = y/x$, and for fixed $t$ define $A(\alpha) = [(1-t) + t\alpha]^4$ and $B(\alpha) = (1-t) + t\alpha^4$.

We note that

$$A(0) = (1-t)^4 < (1-t) = B(0) .$$

Furthermore,

$$A'''(\alpha) = 24t^3 < 24 = B'''(\alpha) .$$

Integrating three times, and applying the fundamental theorem of calculus, we see that $A(\alpha) < B(\alpha)$ and $f$ is strictly convex. ∎

In the converse direction we note that a function with continuous, positive definite Hessian will certainly be strictly convex. We leave the details of that assertion to the reader. [The main thing to note is that $t \mapsto f((1-t)x + ty)$ has positive second derivative for every nonzero choice of $x, y$. This reduces the question to the one-dimensional case, where it is straightforward to treat.]

**Example 2.28** Consider the function $f(x_1, x_2, \ldots, x_N) = |x|^2 \equiv |x_1|^2 + |x_2|^2 + \cdots + |x_N|^2$ on $\mathbb{R}^N$. We can see right away that the real Hessian of $f$ is $2N > 0$, so $f$ is convex from that point of view. Also

$$\begin{aligned} f((1-t)P + tQ) &= [(1-t)P_1 + tQ_1]^2 \\ &\quad + [(1-t)P_2 + tQ_2]^2 \\ &\quad + \cdots + [(1-t)P_N + tQ_N]^2 . \end{aligned}$$

We need to compare this with

$$\begin{aligned} (1-t)f(P) + tf(Q) &= \left[(1-t)P_1^2 + \cdots + (1-t)P_N^2\right] \\ &\quad + \left[tQ_1^2 + \cdots tQ_N^2\right] . \end{aligned}$$

It all comes down to seeing that, for each $j$,

$$\left[(1-t)P_j + tQ_j\right]^2 \le (1-t)P_j^2 + tQ_j^2 ,$$

and that reduces to

$$(t^2 - t)\left[P_j^2 + Q_j^2 - 2P_jQ_j\right] \le 0 .$$

Of course $t^2 - t \le 0$ since $0 \le t \le 1$ and the expression in brackets is nonnegative. So that gives us that the function $f(x) = |x|^2$ is convex from the original definition.

∎

PROPOSITION 2.29 *Let $\| \ \|$ be any norm on $\mathbb{R}^N$. Then $\| \ \|$ is a convex function.*

**Proof:** This is immediate from the triangle inequality. □

The reasoning in our discussion of the real Hessian above can easily be reversed to see that the following is true:

PROPOSITION 2.30 *A $C^2$ function on $\mathbb{R}^N$ is convex if and only if it has positive semi-definite Hessian at each point of its domain.*

The following result is straightforward:

PROPOSITION 2.31 *If $f$ is a convex function then $f$ is continuous.*

**Proof:** See [ZYG] or [VLA, p. 85] for further discussion of these ideas.

Now let $x, y$ lie in the convex domain of $f$. Assume without loss of generality that $f(x) = 0$ and we shall prove that $f$ is continuous at $x$. For $0 \leq t \leq 1$, we know that

$$f((1-t)x + ty) \leq (1-t)f(x) + tf(y) .$$

We think of $t$ as positive and small, so that $(1-t)x + ty$ is close to $x$. Then we see that

$$|f((1-t)x + ty)| \leq t|f(y)| .$$

That shows that

$$\lim_{z \to x} f(z) = 0 = f(x) ,$$

so that $f$ is continuous at $x$. □

Of course it is also useful to consider convex functions on a domain. Certainly we may say that $F : \Omega \to \mathbb{R}$ is convex (with $\Omega$ a convex domain) if

$$\sum_{j,k} \frac{\partial^2 F}{\partial x_j \partial x_k}(x)\alpha_j\alpha_k \geq 0$$

for all $(\alpha_1, \ldots, \alpha_N)$ and all points $x \in \Omega$. Equivalently, $F$ is convex on a convex domain $\Omega$ if, whenever $P, Q \in \Omega$ and $0 \leq \lambda \leq 1$ we have

$$F((1-t)P + tQ) \leq (1-t)f(P) + tf(Q) .$$

Other properties of convex functions are worth noting. For example, if $f : \mathbb{R}^n \to \mathbb{R}$ is convex and $\varphi : \mathbb{R} \to \mathbb{R}$ is convex and increasing then $\varphi \circ f$ is convex. Certainly the sum of any two convex functions is convex.

## 2.4 Exhaustion Functions

It is always useful to be able to characterize geometric properties of domains in terms of functions. For functions are more flexible objects than domains: one can do more with functions. With this thought in mind we make the following definition:

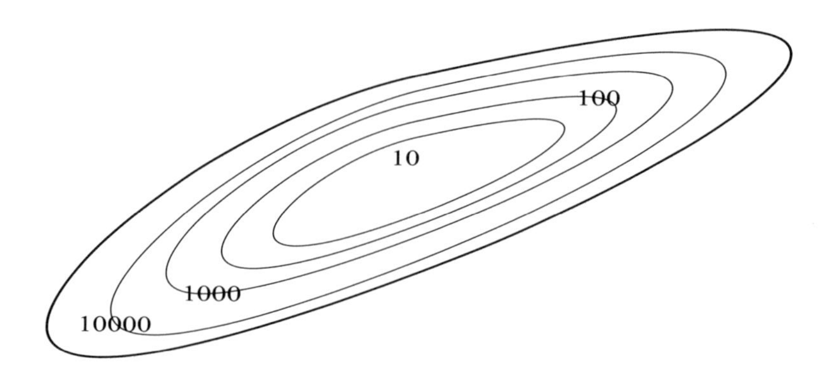


Figure 2.16: A convex exhaustion function.

**Definition 2.32** Let $\Omega \subseteq \mathbb{R}^N$ be a bounded domain. We call a function

$$\lambda : \Omega \to \mathbb{R}$$

an *exhaustion function* if, for each $c \in \mathbb{R}$, the set

$$\lambda^{-1}((-\infty, c]) = \{x \in \Omega : \lambda(x) \leq c\}$$

is a relatively compact subset of $\Omega$ (that is to say, the set has compact closure in the domain). See Figure 2.16.

We are primarily interested in *convex* exhaustion functions. The key idea here is that the function $\lambda$ is real valued, convex, and blows up at the boundary of $\Omega$.

PROPOSITION 2.33 *Let $\Omega$ be the unit ball in $\mathbb{R}^N$. Then the function*

$$\lambda(x) = e^{1/(1-|x|^2)}$$

*is a convex exhaustion function for $\Omega$.*

**Proof:** First we shall discuss the situation, then we shall actually perform some calculations.

Notice that $\lambda$ blows up at the boundary, so it is an exhaustion function.

Next observe that $\varphi(x) = |x|^2$ is certainly (strictly) convex. Thus $1 - |x|^2$ is *not* convex. But then $1/(1 - |x|^2)$ *is* convex.

Of course $e^t$ is a convex, increasing function, so it then follows that $\lambda(x) = e^{1/(1-|x|^2)}$ is convex.

From a calculational point of view, let us restrict attention to dimension one. Then

$$\frac{d}{dx}e^{1/(1-x^2)} = e^{1/(1-x^2)} \cdot \frac{2x}{(1-x^2)^2}$$

and

$$\begin{aligned}\frac{d^2}{dx^2}e^{1/(1-x^2)} &= e^{1/(1-x^2)} \cdot \left[\frac{2x}{(1-x^2)^2}\right]^2 \\ &\quad + e^{1/(1-x^2)} \cdot \frac{2(1-x^2)^2 + 8x^2(1-x^2)}{(1-x^2)^4} .\end{aligned}$$

If we restrict attention to $0 \le x < 1$, then it is clear that the second derivative is positive. □

THEOREM 2.34 *A domain $\Omega \subseteq \mathbb{R}^N$ is convex if and only if it has a continuous, convex exhaustion function.*

**Proof:** If $\Omega$ possesses such an exhaustion function $\lambda$, then the domains

$$\Omega_k \equiv \{x \in \Omega : \lambda < k\}$$

are convex. And $\Omega$ itself is the increasing union of the $\Omega_k$. It follows immediately, from the synthetic definition of convexity, that $\Omega$ is convex. For the converse, observe that if $\Omega$ is convex and $P \in \partial\Omega$, then the tangent hyperplane at $P$ has the form $a \cdot (x - P) = 0$. Here $a$ is a Euclidean unit vector. It then follows that the quantity $a \cdot (x - P)$ is the distance from $x \in \Omega$ to this hyperplane. Now the function

$$\mu_{a,P}(x) \equiv -\log a \cdot (x - P)$$

is convex since one may calculate the Hessian $\mathcal{H}$ directly. Its value at a point $x$ equals

$$\mathcal{H}(b, b) = \frac{(a \cdot b)^2}{[a \cdot (x - P)]^2} \ge 0 .$$

If $\delta_\Omega(x)$ is the Euclidean distance of $x$ to $\partial\Omega$, then

$$-\log \delta_\Omega(x) = \sup_{P \in \partial\Omega} [-\log a \cdot (x - P)] .$$

Thus $-\log \delta_\Omega$ is a convex function that blows up at $\partial\Omega$. Now set

$$\lambda(x) = \max\left\{-\log \delta_\Omega(x), |x|^2\right\} .$$

This is a continuous, convex function that blows up at the boundary. So it is the convex exhaustion function that we seek. □

**Remark 2.35** It is a fact that a domain is convex if and only if it has a $C^\infty$, strictly convex exhaustion function. This information is useful in practice. We prove the result in Theorem 2.40 below.

LEMMA 2.36 *Let $F$ be a convex function on $\mathbb{R}^N$. Then there is a sequences $f_1 \geq f_2 \geq \cdots$ of $C^\infty$, strongly convex functions such that $f_j \to F$ pointwise.*

**Proof:** Let $\varphi$ be a $C_c^\infty$ function which is nonnegative and has integral 1. We may also take $\varphi$ to be supported in the unit ball, and be radial. For $\epsilon > 0$ we set

$$\varphi_\epsilon(X) = \epsilon^{-N}\varphi(x/\epsilon) .$$

We define

$$F_\epsilon(x) = F * \varphi_\epsilon(x) = \int F(x-t)\varphi_\epsilon(t)\, dt .$$

We assert that each $F_\epsilon$ is convex. For let $P, Q \in \mathbb{R}^N$ and $0 \leq \lambda \leq 1$. Then

$$\begin{aligned} & F_\epsilon((1-\lambda)P + \lambda Q) \\ &= \int F((1-\lambda)P + \lambda Q - t)\varphi_\epsilon(t)\, dt \\ &= \int F((1-\lambda)(P-t) + \lambda(Q-t))\varphi_\epsilon(t)\, dt \\ &\leq \int \big[(1-\lambda)F(P-t) + \lambda F(Q-t)\big]\varphi_\epsilon(t)\, dt \\ &= (1-\lambda)F_\epsilon(P) + \lambda F_\epsilon(Q) . \end{aligned}$$

So $F_\epsilon$ is convex.

Now set

$$f_j(x) = F_{\epsilon_j} + \delta_j|x|^2 .$$

Certainly $f_j$ is strongly convex because $F_\epsilon$ is convex and $|x|^2$ strongly convex. If $\epsilon_j > 0$, $\delta_j > 0$ are chosen appropriately, then we will have

$$f_1 \geq f_2 \geq \ldots$$

and $f_j \to F$ pointwise. That is the desired conclusion. □

PROPOSITION 2.37 *Let $F : \mathbb{R}^N \to \mathbb{R}$ be a continuous function. Then $F$ is convex if and only if, for any $\varphi \in C_c^\infty(\mathbb{R}^N)$ with $\varphi \geq 0$, $\int \varphi \, dx = 1$, and any $w = (w_1, w_2, \ldots, w_N) \in \mathbb{R}^N$ it holds that*

$$\int_{\mathbb{R}^N} F(x) \left[ \sum_{j,k} \frac{\partial^2 \varphi}{\partial x_j \partial x_k}(x) w_j w_k \right] dx \geq 0 \,.$$

**Proof:** Assume that $F$ is convex. In the special case that $F \in C^\infty$, we certainly know that

$$\sum_{j,k} \frac{\partial^2 F}{\partial x_j \partial x_k}(x) w_j w_k \geq 0 \,.$$

Hence it follows that

$$\int_{\mathbb{R}^N} \sum_{j,k} \frac{\partial^2 F}{\partial x_j \partial x_k}(x) w_j w_k \cdot \varphi(x) \, dx \geq 0 \,.$$

Now the result follows from integrating by parts twice (the boundary terms vanish since $\varphi$ is compactly supported).

Now the general case follows by approximating $F$ as in the preceding lemma.

For the converse direction, we again first treat the case when $F \in C^\infty$. Assume that

$$\int_{\mathbb{R}^N} \sum_{j,k} \frac{\partial^2 \varphi}{\partial x_j \partial x_k}(x) w_j w_k \cdot F(x) \, dx \geq 0$$

for all suitable $\varphi$. Then integration by parts twice gives us the inequality we want.

For general $F$, let $\psi$ be a nonnegative $C_c^\infty$ function, supported in the unit ball, with integral 1. Set $\psi_\epsilon(x) = \epsilon^{-N} \psi(x/\epsilon)$. Define $F_\epsilon(x) = F * \psi_\epsilon(x)$. Then $F_\epsilon \to F$ pointwise and

$$\int_{\mathbb{R}^N} \sum_{j,k} \frac{\partial^2 \varphi}{\partial x_j \partial x_k}(x) w_j w_k \cdot F(x) \, dx \geq 0$$

certainly implies that

$$\int_{\mathbb{R}^N} \sum_{j,k} \frac{\partial^2 \varphi}{\partial x_j \partial x_k}(x) w_j w_k \cdot F_\epsilon(x) \, dx \geq 0 \,.$$

We may integrate by parts twice in this last expression to obtain

$$\int_{\mathbb{R}^N} \varphi(x) \sum_{j,k} \frac{\partial^2 F_\epsilon}{\partial x_j \partial x_k}(x) w_j w_k \, dx \geq 0 \,.$$

It follows that each $F_\epsilon$ is convex. Thus

$$F_\epsilon((1-\lambda)P + \lambda Q) \le (1-\lambda)F_\epsilon(P) + \lambda F_\epsilon(Q)$$

for every $P, Q, \lambda$. Letting $\epsilon \to 0^+$ yields that

$$F((1-\lambda)P + \lambda Q) \le (1-\lambda)F(P) + \lambda F(Q)$$

hence $F$ is convex. That completes the proof. □

We may think of this last result as characterizing convex functions in the sense of distributions, or in the sense of weak derivatives.

For applications in the next theorem, it is useful to note a new result. We first give a definition.

**Definition 2.38** A twice continuously differentiable function $u$ on a domain $\Omega$ is *subharmonic* if $\triangle u \ge 0$ at all points.[4] A function that is not twice differentiable is subharmonic if it is the monotone limit of twice continuously differentiable subharmonic functions.

PROPOSITION 2.39 *Any convex function $f$ is subharmonic.*

**Proof:** To see this, let $P$ and $P'$ be distinct points in the domain of $f$ and let $X$ be their midpoint. Then certainly

$$2f(X) \le f(P) + f(P') .$$

Let $\eta$ be any special orthogonal rotation centered at $X$. We may write

$$2f(X) \le f(\eta(P)) + f(\eta(P')) .$$

Now average over the special orthogonal group to derive the usual sub-mean-value property for subharmonic functions. □

The last topic is also treated quite elegantly in Chapter 3 of [HOR]. One may note that the condition that the Hessian be positive semi-definite is stronger than the condition that the Laplacian be nonnegative—this just by simple algebra. That gives another proof of our result.

[4] A key property of subharmonic functions is the sub-mean value property: If $B(x,r)$ is a Euclidean ball in the domain of the function, then

$$f(x) \le \frac{1}{|B(x,r)|} \int_{B(x,r)} f(t)\, dt .$$

See the Appendix.

THEOREM 2.40 *A domain $\Omega \subseteq \mathbb{R}^N$ is convex if and only if it has a $C^\infty$, strictly convex exhaustion function.*

**Proof:** Only the forward direction need be proved (as the converse direction is contained in the last theorem).

We build the function up iteratively. We know by the preceding theorem that there is a continuous exhaustion function $\lambda$. Let

$$\Omega_c = \{x \in \Omega : \lambda(x) + |x|^2 < c\}$$

for $c \in \mathbb{R}$. Then each $\Omega_c \subseteq \Omega$ and $c' > c$ implies that $\Omega_c \subseteq \Omega'_c$. Now let $0 \leq \varphi \in C_c^\infty(\mathbb{R}^N)$ with $\int \varphi \, dx = 1$, $\varphi$ radial.[5] We may take $\varphi$ to be supported in $B(0,1)$. Let $0 < \epsilon_j < \operatorname{dist}(\Omega_{j+1}, \partial\Omega)$. If $x \in \Omega_{j+1}$, set

$$\lambda_j(x) = \int_\Omega [\lambda(t) + |t|^2]\epsilon_j^{-N}\varphi((x-t)/\epsilon_j)\, dV(t) + |x|^2 + 1\,.$$

Then each $\lambda_j$ is $C^\infty$ and strictly convex on $\Omega_{j+1}$. Moreover, by the previously noted subharmonicity of $\lambda$, we may be sure that $\lambda_j(x) > \lambda(x) + |x|^2$ on $\overline{\Omega}_j$.

Now let $\chi \in C^\infty(\mathbb{R})$ be a convex function with $\chi(t) = 0$ for $t \leq 0$ and $\chi'(t), \chi''(t) > 0$ when $t > 0$. Note that, $\Psi_j(x) \equiv \chi(\lambda_j(x) - (j-1))$ is positive and convex on $\Omega_j \setminus \overline{\Omega}_{j-1}$ and is, of course, $C^\infty$. Notice now that $\lambda_0 > \lambda$ on $\Omega_0$. If $a_1$ is large and positive, then $\lambda'_1 \equiv \lambda_0 + a_1\Psi_1 > \lambda$ on $\Omega_1$. Inductively, if $a_1, a_2, \ldots a_{\ell-1}$ have been chosen, select $a_\ell > 0$ such that $\lambda'_\ell \equiv \lambda_0 + \sum_{j=1}^{\ell} a_j\Psi_j > \lambda$ on $\Omega_\ell$.

Since $\Psi_{\ell+k} = 0$ on $\Omega_\ell$, $k > 0$, we see that $\lambda'_{\ell+k} = \lambda'_{\ell+k'}$ on $\Omega_\ell$ for any $k, k' > 0$. So the sequence $\lambda'_\ell$ stabilizes on compacta and $\lambda' \equiv \lim_{\ell\to\infty} \lambda'_\ell$ is a $C^\infty$ strictly convex function that majorizes $\lambda$. Hence $\lambda'$ is the smooth, strictly convex exhaustion function that we seek. □

COROLLARY 2.41 *Let $\Omega \subseteq \mathbb{R}^N$ be any convex domain. Then we may write*

$$\Omega = \bigcup_{j=1}^{\infty} \Omega_j\,,$$

*where this is an increasing union and each $\Omega_j$ is strongly analytically convex with $C^\infty$ boundary.*

[5]Here a function $\varphi$ is radial if $\varphi(a) = \varphi(b)$ whenever $|a| = |b|$. In other words, the function should have circular symmetry. See the Appendix.

**Proof:** Let $\lambda$ be a smooth, strictly convex exhaustion function for $\Omega$. By Sard's theorem (see [KRP1]),[6] there is a strictly increasing sequence of values $c_j \to +\infty$ so that each

$$\Omega_{c_j} = \{x \in \Omega : \lambda(x) < c_j\}$$

has smooth boundary. Then of course each $\Omega_{c_j}$ is strongly analytically convex. And the $\Omega_{c_j}$ form an increasing sequence of domains whose union is $\Omega$.

# Exercises

**1.** Is the domain

$$\Omega = \{(x_1, x_2) \in \mathbb{R}^2 : |x_1|^2 + |x_2|^4 < 1\}$$

strongly analytically convex?

**2.** Is the domain

$$\Omega = \{(x_1, x_2) \in \mathbb{R}^2 : |x_1|^4 + |x_2|^4 < 1\}$$

strongly analytically convex?

**3.** Show that the function $f : \mathbb{R} \to \mathbb{R}$ given by

$$f(x) = \begin{cases} e^{-1/x^2} & \text{if} \quad x > 0 \\ 0 & \text{if} \quad x \leq 0 \end{cases}$$

is weakly analytically convex.

**4.** Show that the function

$$f(x) = e^{x^2}$$

is strongly analytically convex.

**5.** Prove this version of the Weierstrass approximation theorem.

> Let $f$ be a twice continuously differentiable, convex function on the interval $I = [0, 1]$. Then $f$ may be uniformly approximated by a convex polynomial function.

[**Hint:** Apply the classical Weierstrass theorem to the second derivative of $f$.]

**6.** Prove that the function $y = \ln x$ is strongly concave.

---

[6] Sard's theorem says that the set of singular values of a smooth function is thin. See the Appendix and also [KRP1].

**7.** Prove that the function $y = \sqrt{x}$ is strongly concave.

**8.** Prove that a convex function of one real variable can be approximated by piecewise linear, convex functions.

**9.** If $\varphi : \mathbb{R} \to \mathbb{R}$ is nonnegative and continuous, then show that the function obtained by antidifferentiating $\varphi$ twice will be convex. What additional condition on $\varphi$ will guarantee that the second antiderivative is strongly convex?

**10.** What are necessary and sufficient conditions on $a, b, c$ to guarantee that $f(x) = ax^2 + bx + c$ is strongly convex? Weakly convex?

**11.** What are necessary and sufficient conditions on $a, b, c, d$ to guarantee that $f(x) = ax^3 + bx^2 + cx + d$ is strongly convex? Weakly convex?

**12.** Write down a defining function for the planar triangle with vertices $(\pm 1, 0)$ and $(0, 1)$.

**13.** Calculate the Hessian of the defining function for the domain

$$\{(x_1, x_2) \in \mathbb{R}^2 : x_1^4 + x_2^4 < 1\} .$$

Which boundary points are strongly analytically convex? Which boundary points are weakly analytically convex?

**14.** Write down a $C^\infty$, strictly convex exhaustion function for the domain

$$\{(x_1, x_2) \in \mathbb{R}^2 : x_1^2 + x_2^4 < 1\} .$$

**15.** Give an example of a sequence of strongly convex functions $f_j$ on $\mathbb{R}$ that converge to a function $f$ that is weakly convex.

# Chapter 3

# Further Developments Using Functions

**Prologue:** In the present chapter we introduce a notion of convexity with respect to a family of functions. This idea can be used to study not only traditional convex domains but also other families of domains. It is a flexible tool, and one with many applications.

We go on to consider a ranked ordering of different degrees of convexity. Again, this idea has no precedent in the classical theory of Chapter 1. It is a fine tool for differentiating sets.

We next examine extreme points and set the stage for our later study of the very important Krein-Milman theorem. This result is one of the hallmarks of convexity theory, and there is much to say about it.

Our next study is of support functions—yet another way to bring the theory of functions to bear on our convexity studies. This book demonstrates in decisive ways that functions are very flexible tools.

We return to the idea of approximation and consider affine convex functions and their approximation properties. This is hard analysis at its best.

We conclude the chapter with a treatment of the idea of bumping. This is a perturbation theory for convex domains. It is of interest, for example, in control theory.

## 3.1 Other Characterizations of Convexity

Let $\Omega \subseteq \mathbb{R}^N$ be a domain and let $\mathcal{F}$ be a family of real-valued functions on $\Omega$ (we do not assume in advance that $\mathcal{F}$ is closed under any algebraic operations, although often in practice it will be). Let $K$ be a compact subset of $\Omega$. Then the *convex hull of* $K$ *in* $\Omega$ *with respect to* $\mathcal{F}$ is defined to be

$$\widehat{K}_{\mathcal{F}} \equiv \left\{ x \in \Omega : f(x) \leq \sup_{t \in K} f(t) \text{ for all } f \in \mathcal{F} \right\}.$$

We sometimes denote this hull by $\widehat{K}$ when the family $\mathcal{F}$ is understood or when no confusion is possible. We say that $\Omega$ is *convex* with respect to $\mathcal{F}$ provided $\widehat{K}_{\mathcal{F}}$ is compact in $\Omega$ whenever $K$ is. [When the functions in $\mathcal{F}$ are complex-valued then $|f|$ replaces $f$ in the definition of $\widehat{K}_{\mathcal{F}}$.]

PROPOSITION 3.1 *Let* $\Omega \subseteq \mathbb{R}^N$ *and let* $\mathcal{F}$ *be the family of real linear functions. Then* $\Omega$ *is convex with respect to* $\mathcal{F}$ *if and only if* $\Omega$ *is geometrically convex.*

**Proof:** First suppose that $\Omega$ is convex with respect to the family $\mathcal{F}$. Let $P$ and $Q$ be points of $\Omega$. Then the convex hull of $\{P, Q\}$ with respect to $\mathcal{F}$ certainly contains the interval $\overline{PQ}$ (that is to say, a linear function $\varphi$ will take all values between $\varphi(P)$ and $\varphi(Q)$ on that segment). So the segment will be relatively compact in $\Omega$, hence will be contained in $\Omega$.

Conversely, assume that $\Omega$ is convex according to the classical definition on the first page of this book. Let $K$ be a compact subset of $\Omega$. Then the closed convex hull of $K$ will be compact in $\Omega$. But that hull is nothing other than the hull of $K$ with respect to $\mathcal{F}$. Hence $\Omega$ is convex with respect to $\mathcal{F}$. $\square$

PROPOSITION 3.2 *Let* $\Omega \subseteq \mathbb{R}^N$ *be any domain. Let* $\mathcal{F}$ *be the family of continuous functions on* $\Omega$. *Then* $\Omega$ *is convex with respect to* $\mathcal{F}$.

**Proof:** If $K \subseteq \Omega$ and $x \notin K$ then the function $F(t) = 1/(1 + |x - t|)$ is continuous on $\Omega$. Notice that $f(x) = 1$ and $|f(k)| < 1$ for all $k \in K$. Thus $x \notin \widehat{K}_{\mathcal{F}}$. Therefore $\widehat{K}_{\mathcal{F}} = K$ and $\Omega$ is convex with respect to $\mathcal{F}$. $\square$

What the last two results are telling us is that the family of linear functions is rather rigid, while the family of continuous functions is "flabby" (in the

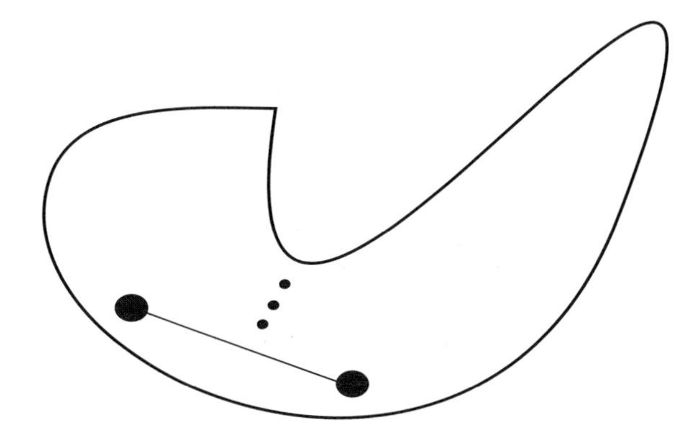

Figure 3.1: A domain that fails the segment characterization.

sense of sheaf theory). Put in other words, continuous functions are easily bent while linear functions are not.

We close this discussion of convexity with a geometric characterization of the property. We shall, later in the book, refer to this as the "segment characterization." First, if $\Omega \subseteq \mathbb{R}^N$ is a domain and $I$ is a closed one-dimensional segment lying in $\Omega$, then the boundary $\partial I$ is the set consisting of the two endpoints of $I$.

PROPOSITION 3.3 *The domain $\Omega$ is convex if and only if, whenever $\{I_j\}_{j=1}^\infty$ is a collection of closed segments in $\Omega$ and $\{\partial I_j\}$ is relatively compact in $\Omega$, then so is $\{I_j\}$. See Figure 3.1.*

**Remark 3.4** This is little more than a restatement of the classical definition of geometric convexity. The proof will make this clear.

**Proof of the Proposition:** Suppose that $\Omega$ satisfies the segment characterization. Let

$$\mathcal{S} = \{(P, Q) : P \in \Omega, Q \in \Omega, \text{and } \overline{PQ} \subseteq \Omega\}.$$

It is clear that $\mathcal{S}$ is not empty, for we may take $P$ to be any point of $\Omega$ and then let $Q = P$. We claim also that $\mathcal{S}$ is both open and closed in the relative topology of $\Omega$.

That $\mathcal{S}$ is open is obvious, for $\Omega$ is an open set. If $P, Q \in \Omega$ and $\overline{PQ} \subseteq \Omega$, then points near enough to $P$ and $Q$ are also in $\Omega$ and segments near to $\overline{PQ}$ also lie in $\Omega$.

For closedness, suppose that $(P_j, Q_j) \in \Omega$ for $j = 1, 2, \ldots$ and assume that $P_j \to P_0 \in \Omega$ and $Q_j \to Q_0 \in \Omega$ (remember that we are considering the relative topology). Then $\{P_j\}$ is relatively compact in $\Omega$ and $\{Q_j\}$ is relatively compact in $\Omega$. It follows then that the set of segments $\overline{P_jQ_j}$ is relatively compact in $\Omega$, hence these segments converge to a limit segment that lies in $\Omega$. Of course that limit segment is $\overline{P_0Q_0}$.

For the converse, suppose that $\Omega$ is a domain that is convex according to the classical definition. If $\{I_j\}$ is a collection of segments in $\Omega$ whose endpoints form a relatively compact set, then the segments $I_j$ themselves of course lie in $\Omega$. If they do not form a relatively compact set, then there is a limit segment that does not lie entirely in $\Omega$. But that limit segment will have endpoints that do lie in $\Omega$. That contradicts the classical definition of convexity. □

In fact the statement of the proposition admits of many variants. One of these is the following:

PROPOSITION 3.5 *Let $\Omega \subseteq \mathbb{R}^N$ be convex. If $\{I_j\}$ is a collection of closed segmenets in $\Omega$ then*

$$dist\,(\partial I_j, \partial\Omega)$$

*is bounded from 0 if and only if*

$$dist\,(I_j, \partial\Omega)$$

*is bounded from 0.*

The following example puts these ideas in perspective.

**Example 3.6** Let $\Omega \subseteq \mathbb{R}^2$ be

$$\Omega = B((0,0),2) \setminus \overline{B((2,0),1)}\,.$$

See Figure 3.2. Let

$$I_j = \{(1-1/j, t) : -1/2 \le t \le 1/2\}\,.$$

Then it is clear that

$$\{\partial I_j\}$$

is relatively compact in $\Omega$ while

$$\{I_j\}$$

is not. And of course $\Omega$ is not convex. ∎

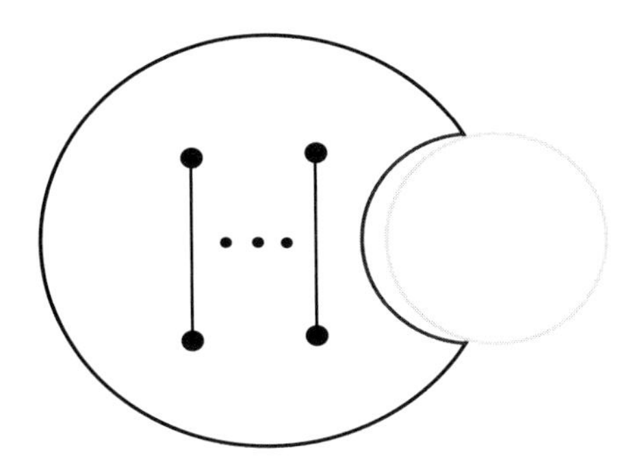

Figure 3.2: A nonconvex domain.

**Example 3.7** By contrast, consider the domain

$$\Omega = \{(x_1, x_2) \in \mathbb{R}^2 : |x_1|^2 + |x_2|^2 < 1\} .$$

This is of course the unit disc.

If $I$ is any closed segment lying in $\Omega$, then it is a simple fact that the distance of the interior of $I$ to the boundary is greater than or equal to the distance of the endpoints to the boundary. The segment property immediately follows. ∎

## 3.2 Convexity of Finite Order

There is a fundamental difference between the domains

$$B = \{x = (x_1, x_2) \in \mathbb{R}^2 : x_1^2 + x_2^2 < 1\}$$

and

$$E = \{x = (x_1, x_2) \in \mathbb{R}^2 : x_1^4 + x_2^4 < 1\} .$$

Both of these domains are convex. The first of these is strongly analytically convex and the second is not. See Figure 3.3. More generally, each of the domains

$$E_m = \{x = (x_1, x_2) \in \mathbb{R}^2 : x_1^{2m} + x_2^{2m} < 1\}$$

is, for $m = 2, 3, \ldots$, weakly (not strongly) analytically convex. Somehow the intuition is that, as $m$ increases, the domain $E_m$ becomes *more* weakly analytically convex. Put differently, the boundary points $(\pm 1, 0)$ are becoming

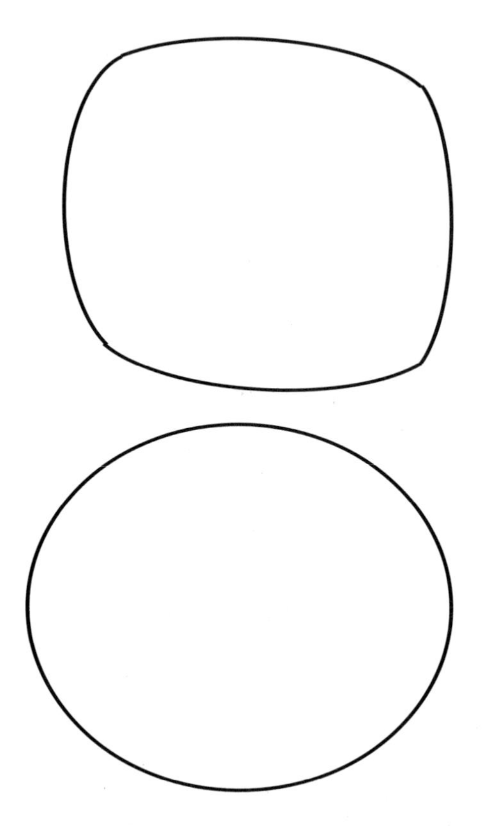

Figure 3.3: Different orders of convexity.

flatter and flatter as $m$ increases. In fact it is interesting to note that, as $m \to \infty$, the domains $E_m$ tend (in the Hausdorff metric on sets, for instance) to the solid unit square. See the Appendix for this concept.

We would like to have a way of quantifying, indeed of measuring, the indicated flatness. These considerations lead to a new definition. We first need a bit of terminology.

Let $f$ be a function on an open set $U \subseteq \mathbb{R}^N$ and let $P \in \Omega$. We say that *$f$ vanishes to order $k$ at $P$* if any derivative of $f$, up to and including order $k$, vanishes at $P$. Thus if $f(P) = 0$ but $\nabla f(P) \neq 0$ then we say that $f$ vanishes to order 0. If $f(P) = 0$, $\nabla f(P) = 0$, $\nabla^2 f(P) = 0$, and $\nabla^3 f(P) \neq 0$, then we say that $f$ vanishes to order 2. See the Appendix for further details. Also refer to Figure 3.4.

Let $\Omega$ be a domain and $P \in \partial\Omega$. Suppose that $\partial\Omega$ is smooth near $P$. We say that the tangent plane $T_P(\partial\Omega)$ has order of contact $k$ with $\partial\Omega$ at $P$ if the

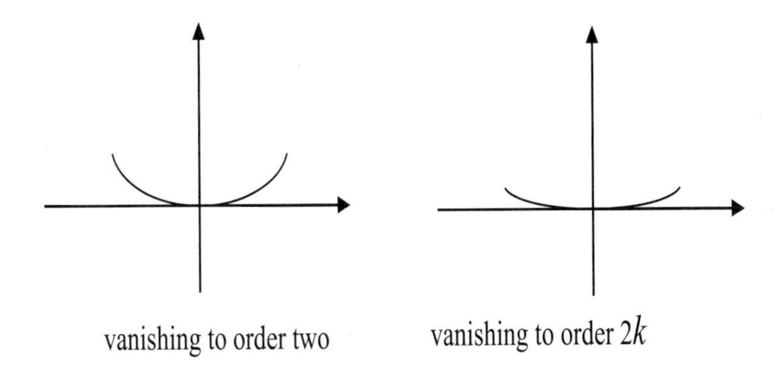


Figure 3.4: Vanishing to order $k$.

defining function $\rho$ for $\Omega$ satisfies

$$|\rho(x)| \leq C|x - P|^k \qquad \text{for all } x \in T_P(\partial\Omega)\,,$$

and this same inequality does *not* hold with $k$ replaced by $k + 1$.

**Definition 3.8** Let $\Omega \subseteq \mathbb{R}^N$ be a domain and $P \in \partial\Omega$ a point at which the boundary is at least $C^k$ for $k$ a positive integer. We say that $P$ is *analytically convex of order* $k$ if

- The point $P$ is convex;
- The tangent plane to $\partial\Omega$ at $P$ has order of contact $k$ with the boundary at $P$.

A domain is convex of order $k$ if each boundary point is convex of order at most $k$, and some boundary point is convex of order exactly $k$.

**Example 3.9** Notice that a point of strong convexity will be analytically convex of order 2. The boundary point $(1, 0)$ of the domain

$$E_{2k} = \{(x_1, x_2) \in \mathbb{R}^2 : x_1^2 + x_2^{2k} < 1\}$$

is analytically convex of order $2k$.

The domain

$$E_{2\infty} = \{(x_1, x_2) \in \mathbb{R}^2 : x_1^2 + e^{-1/|x_2|^2} < 1\}$$

is not convex of any positive order at the boundary points $(1, 0)$ and $(-1, 0)$. We sometimes say that these boundary points are convex of *infinite order*. ∎

PROPOSITION 3.10 *Let $\Omega \subseteq \mathbb{R}^N$ be a bounded domain, and let $P \in \partial\Omega$ be convex of finite order. Then that order is an even number.*

**Proof:** Let $m$ be the order of the point $P$.

We may assume that $P$ is the origin and that the outward normal direction at $P$ is the $x_1$ direction. If $\rho$ is a defining function for $\Omega$ near $P$ then we may use the Taylor expansion about $P$ to write

$$\rho(x) = 2x_1 + \varphi(x),$$

and $\varphi$ will vanish to order $m$. If $m$ is odd, then the domain will not lie on one side of the tangent hyperplane

$$T_P(\partial\Omega) = \{x : x_1 = 0\}.$$

So $\Omega$ cannot be convex. □

A very important feature of convexity of finite order is its stability. We formulate that property as follows:

PROPOSITION 3.11 *Let $\Omega \subseteq \mathbb{R}^N$ be a smoothly bounded domain and let $P \in \partial\Omega$ be a point that is convex of finite order $m$. Then points in $\partial\Omega$ that are sufficiently near $P$ are also convex of finite order at most $m$.*

**Proof:** Let $\Omega = \{x \in \mathbb{R}^N : \rho(x) < 0\}$, where $\rho$ is a defining function for $\Omega$. Then the "finite order" condition is given by the nonvanishing of a derivative of $\rho$ at $P$. Of course that same derivative will be nonvanishing at nearby points, and that proves the result. □

PROPOSITION 3.12 *Let $\Omega \subseteq \mathbb{R}^N$ be a smoothly bounded domain. Then there will be a point $P \in \partial\Omega$ and a neighborhood $U$ of $P$ so that each point of $U \cap \partial\Omega$ will be analytically convex of order 2 (i.e., strongly analytically convex).*

This repeats, in slightly different language, a result that we presented earlier in a different context.

**Proof:** Let $D$ be the diameter of $\Omega$. We may assume that $\overline{\Omega}$ is distance at least $10D + 10$ from the origin 0. Let $P$ be the point of $\partial\Omega$ which is furthest (in the Euclidean metric) from 0. Then $P$ is the point that we seek. Refer to Figure 3.5.

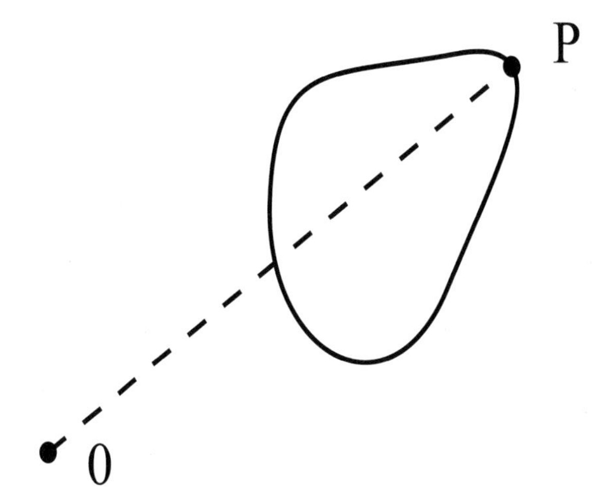


Figure 3.5: A boundary point of strong convexity.

Let $L$ be the distance of 0 to $P$. Then we see that the sphere with center 0 and radius $L$ externally osculates $\partial\Omega$ at $P$. Of course the sphere is strongly analytically convex at the point of contact. Hence so is $\partial\Omega$. By the continuity of second derivatives of the defining function for $\Omega$, the same property holds for nearby points in the boundary. That completes the proof. □

**Example 3.13** Consider the domain

$$\Omega = \{(x_1, x_2, x_3) \in \mathbb{R}^3 : x_1^2 + x_2^4 + x_3^4 < 1\}.$$

The boundary points of the form $(0, a, b)$ are analytically convex of order 4. All others are analytically convex of order 2 (i.e., strongly analytically convex). ∎

It is straightforward to check that Euclidean isometries preserve convexity, preserve strong convexity, and preserve convexity of finite order. Diffeomorphisms do not. In fact we have:

PROPOSITION 3.14 *Let $\Omega_1$, $\Omega_2$ be smoothly bounded domains in $\mathbb{R}^N$, let $P_1 \in \partial\Omega_1$ and $P_2 \in \partial\Omega_2$. Let $\Phi$ be a diffeomorphism from $\overline{\Omega_1}$ to $\overline{\Omega_2}$ and assume that $\Phi(P_1) = P_2$. Further suppose that the Jacobian matrix of $\Phi$ at $P_1$ is an orthogonal linear mapping.*[1] *Then we have:*

[1] A mapping is *orthogonal* if it preserves the lengths of vectors and preserves the Euclidean inner product. See the Appendix.

- *If $P_1$ is a convex boundary point, then $P_2$ is a convex boundary point;*
- *If $P_1$ is a strongly analytically convex boundary point, then $P_2$ is a strongly analytically convex boundary point;*
- *If $P_1$ is a boundary point that is analytically convex of order $2k$, then $P_2$ is a boundary point that is analytically convex of order $2k$.*

**Proof:** We consider the first assertion. Let $\rho$ be a defining function for $\Omega_1$. Then $\rho \circ \Phi^{-1}$ will be a defining function for $\Omega_2$. Of course we know that the Hessian of $\rho$ at $P_1$ is positive semi-definite. It is straightforward to calculate the Hessian of $\widetilde{\rho} \equiv \rho \circ \Phi^{-1}$ and see that it is just the Hessian of $\rho$ composed with $\Phi$ applied to the vectors transformed under $\Phi^{-1}$. So of course $\widetilde{\rho}$ will have positive semi-definite Hessian.

The other two results are verified using the same calculation. □

PROPOSITION 3.15 *Let $\Omega$ be a smoothly bounded domain in $\mathbb{R}^N$. Let $\mathcal{L}$ be an invertible linear map on $\mathbb{R}^N$. Define $\Omega' = \mathcal{L}(\Omega)$. Then*

- *Each convex boundary point of $\Omega$ is mapped to a convex boundary point of $\Omega'$.*
- *Each strongly analytically convex boundary point of $\Omega$ is mapped to a strongly analytically convex boundary point of $\Omega'$.*
- *Each boundary point of $\Omega$ that is analytically convex of order $2k$ is mapped to a boundary point of $\Omega'$ that is analytically convex of order $2k$.*

**Proof:** Obvious. □

Maps which are not invertible tend to decrease the order of a convex point. An example will illustrate this idea:

**Example 3.16** Let $\Omega = \{(x_1, x_2) \in \mathbb{R}^2 : x_1^2 + x_2^2 < 1\}$ be the unit ball and $\Omega' = \{(x_1, x_2) \in \mathbb{R}^2 : x_1^4 + x_2^4 < 1\}$. We see that

$$\Phi(x_1, x_2) = (x_1^2, x_2^2)$$

maps $\Omega'$ onto $\Omega$. And we see that $\Omega$ is strongly analytically convex (i.e., analytically convex of order 2 at each boundary point) while $\Omega'$ has boundary points that are analytically convex of order 4. The analytically convex points of order 4 are mapped by $\Phi$ to analytically convex points of order 2. ∎

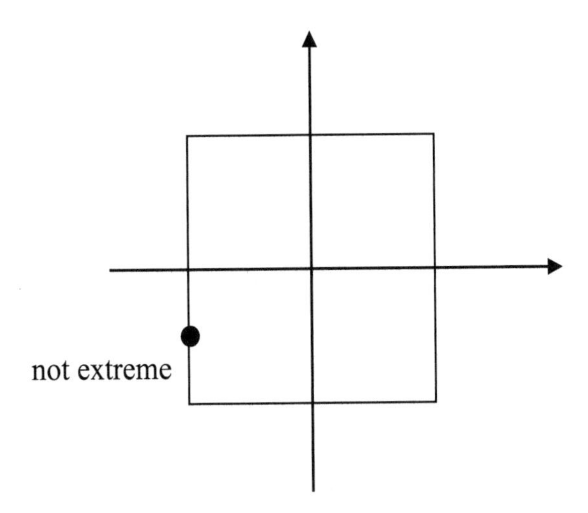


Figure 3.6: A convex point that is not extreme.

## 3.3 Extreme Points

A point $P \in \partial\Omega$ is called an *extreme point* if, whenever $a, b \in \partial\Omega$ and $P = (1-\lambda)a + \lambda b$ for some $0 \le \lambda \le 1$ then $a = b = P$.

It is easy to see that, on a convex domain, a point of strong convexity must be extreme, and a point that is analytically convex of order $2k$ must be extreme. But convex points in general are *not* extreme.

**Example 3.17** Let

$$\Omega = \{(x_1, x_2) \in \mathbb{R}^2 : |x_1| < 1, |x_2| < 1\} .$$

Then $\Omega$ is clearly convex. But any boundary point with $x_1$, $x_2$ not both 1 is not extreme.

For example, consider the boundary point $(-1, -1/2)$. Then

$$(-1, -1/2) = \frac{1}{2}(-1, -1) + \frac{1}{2}(-1, 0) .$$

See Figure 3.6. ∎

**Example 3.18** Let $\Omega \subseteq \mathbb{R}^2$ be the domain with boundary consisting of

- The segments from $(-3/4, 1)$ to $(3/4, 1)$, from $(1, 3/4)$ to $(1, -3/4)$, from $(3/4, -1)$ to $(-3/4, -1)$, and from $(-1, 3/4)$ to $(-1, -3/4)$.

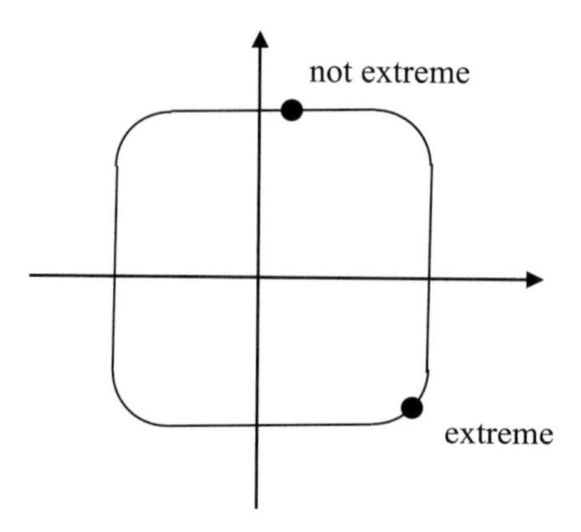


Figure 3.7: Extreme points and non-extreme points.

- The four circular arcs

$$(x+3/4)^2 + (y-3/4)^2 = \frac{1}{16}\,, \quad y \ge 0\,,\ x \le 0\,;$$

$$(x-3/4)^2 + (y-3/4)^2 = \frac{1}{16}\,, \quad y \ge 0\,,\ x \ge 0\,;$$

$$(x-3/4)^2 + (y+3/4)^2 = \frac{1}{16}\,, \quad y \le 0\,,\ x \ge 0\,;$$

$$(x+3/4)^2 + (y+3/4)^2 = \frac{1}{16}\,, \quad y \le 0\,,\ x \le 0\,;$$

See Figure 3.7. Then any point on any of the circular arcs is extreme. But no other boundary point is extreme. Note, however, that the extreme points $(-3/4, 1)$, $(3/4, 1)$, $(1, -3/4)$, $(1, 3/4)$, $(-3/4, -1)$, $(3/4, -1)$, $(-1, -3, 4)$, and $(-1, 3/4)$ are *not* convex of finite order. Refer again to Figure 3.7. ∎

PROPOSITION 3.19 *Let $\Omega \subseteq \mathbb{R}^N$ be convex and smoothly bounded. Let $P \in \partial\Omega$ be analytically convex of order $k$, $2 \le k < \infty$. Then $P$ is an extreme point.*

**Proof:** For convenience we restrict attention to dimension 2. We may assume that the tangent line at $P$ is the $x$-axis, and that the domain lies in the upper halfplane.

Examine the definition of extreme point. First restrict attention to $a$ and $b$ very near to $P$. Then it is clear from the definition of convex to order $k$ that both $a$ and $b$ lie above the $x$-axis. So it cannot be that $P = (1-t)a + tb$. So $P$ is extreme.

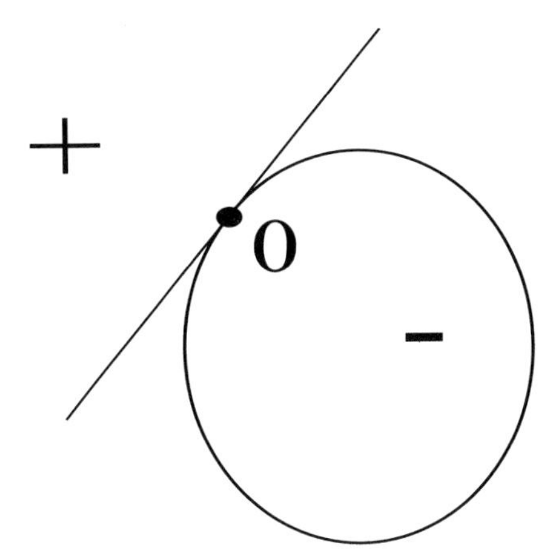


Figure 3.8: A support function.

If instead $a$ and $b$ are far from $P$, then convexity implies that $a$ and $b$ are even further up in the upper halfplane. So, again, $P$ is extreme. $\square$

## 3.4 Support Functions

Let $\Omega \subseteq \mathbb{R}^N$ be a bounded, convex domain with $C^2$ boundary. If $P \in \partial\Omega$ then let $T_P(\partial\Omega)$ be the tangent hyperplane to $\partial\Omega$ at $P$. We may take the outward unit normal at $P$ to be the positive $x_1$ direction. Then the function

$$L(x) = x_1$$

is a linear function that is negative on $\Omega$ and positive on the other side of $T_P(\Omega)$. The function $L$ is called a *support function* for $\Omega$ at $P$. Note that if we take the infimum of all support functions for $P \in \partial\Omega$, normalized so that $|\nabla L(P)| = 1$, then we obtain a defining function for $\Omega$. Refer to Figure 3.8.

The support function of course takes the value 0 at $P$. It may take the value 0 at other boundary points—for instance in the case of the domain $\{(x_1, x_2) : |x_1| < 1, |x_2| < 1\}$. But if $\Omega$ is convex and $P \in \partial\Omega$ is a point of convexity of finite order $2k$, then the support function will vanish on $\partial\Omega$ only at the point $P$. The same assertion holds when $P$ is an extreme point of the boundary.

## 3.5 Approximation from Below by Affine Functions

A useful piece of terminology is the following. Let $f : \mathbb{R}^N \to \mathbb{R}$ be a function. Then the *epigraph* of $f$, denote by $\mathcal{E}(f)$ or simply by $\mathcal{E}$ when no confusion is possible, is the set $\{(x, y) : x \in \mathbb{R}^N, y \geq f(x)\}$. It is clear that the function $f$ is convex if and only if the epigraph is a convex set. We shall make good use of the epigraph in the results to be presented below.

We will also make use of affine functions. A function $\varphi : \mathbb{R}^N \to \mathbb{R}$ is *affine* if $\varphi(x) = b \cdot x + a$ for some $b \in \mathbb{R}^N$ and some $a \in \mathbb{R}$. An affine function is a linear function that has been (possibly) translated. See the Appendix for more on affine mappings.

We acknowledge [TUY] as the source of this exposition.

THEOREM 3.20 *A convex function $f$ on $\mathbb{R}^N$ is the upper envelope (i.e., pointwise supremum) of the family of all affine functions $h$ on $\mathbb{R}^N$ which lie below $f$.*

**Proof:** Let $(a, b) \notin \mathcal{E}(f)$. We claim that there is a point $(c, \gamma) \in \mathbb{R}^N \times \mathbb{R}$ such that

$$\langle c, x\rangle - t < \gamma < \langle c, a\rangle - b \quad \forall (x, t) \in \mathcal{E}(f)\,. \tag{3.20.1}$$

In fact, since $\mathcal{E}$ is a closed convex set then there is a hyperplane separating $(a, b)$ from $\mathcal{E}$. Expressed analytically, there is an affine function $\varphi(x, t) = \langle c, x\rangle + \rho t$ such that

$$\langle c, x\rangle + \rho t < \gamma < \langle c, a\rangle + \rho b \quad \forall (x, t) \in \mathcal{E}\,.$$

Of course $\rho \leq 0$; because, if $\rho > 0$ then, by taking a point $\overline{x}$ in $\mathbb{R}^N$ and an arbitrary $t \geq f(\overline{x})$, we would have $(\overline{x}, t) \in \mathcal{E}$. Thus $\langle c, \overline{x}\rangle + \rho t < \gamma$ for all $t \geq f(\overline{x})$. This would yield a contradiction as $t \to +\infty$. Further note that, if $a \in \mathbb{R}^N$, then $\rho = 0$ would imply that $\langle c, a\rangle < \langle c, a\rangle$, which is impossible. So we certainly know that $\rho < 0$; we can then divide $c$ and $\gamma$ by $-\rho$ and can thus assume that $\rho = -1$. Hence (3.20.1) is valid.

Notice that (3.20.1) implies that $\langle c, x\rangle - \gamma \leq f(x)$ for all $x$. Thus the affine function $h(x) = \langle c, x\rangle - \gamma$ minorizes $f$. Let $\mathcal{Q}$ be the family of all affine functions that minorize $f$. We claim that

$$f(x) = \sup\{h(x) : h \in \mathcal{Q}\}\,. \tag{3.20.2}$$

Suppose to the contrary that $f(a) > \mu \equiv \sup\{h(x) : h \in \mathcal{Q}\}$ for some point $a$. Then $(a, \mu) \notin \mathcal{E}$ and, by what went before, there is a point $(c, \gamma) \in \mathbb{R}^N \times \mathbb{R}$ such that (3.20.1) holds for $b = \mu$. Hence $h(x) = \langle c, x\rangle - \gamma \in \mathcal{Q}$ and $\gamma < \langle c, a\rangle - \mu$; that is to say, $h(a) = \langle c, a\rangle - \gamma > \mu$ and that is a contradiction. So (3.20.2) holds and we are done. □

**Definition 3.21** Let $y = f(x)$ be a function on $\mathbb{R}$. The *convex hull* of $f$ is the function whose epigraph is the convex hull of the epigraph $\mathcal{E}(f)$ of $f$.

COROLLARY 3.22 *For any function $f : \mathbb{R}^N \to \mathbb{R}$, the closure of the convex hull of $f$ is equal to the upper envelope of all affine functions minorizing $f$.*

**Proof:** An affine function $h$ minorizes $f$ if and only if it minorizes the closure of the convex hull of $f$. That gives the result. □

PROPOSITION 3.23 *Any convex function $f : \mathbb{R}^N \to \mathbb{R}$ has an affine minorant. In fact, if $x^0 \in \mathbb{R}^N$, then there exists an affine minorant $h$ which is "exact" at $x^0$; that is to say, so that $h(x^0) = f(x^0)$.*

**Proof:** We use notation from the proof of Theorem 3.20. If $x^0 \in \mathbb{R}^N$, then the point $(x^0, f(x^0))$ lies on the boundary of the epigraph $\mathcal{E}$. Any tangent hyperplane is then a supporting hyperplane. In other words, there is a point $(c, \gamma) \in \mathbb{R}^N \times \mathbb{R}$ so that either $c \neq 0$ or $\gamma \neq 0$ and the affine function

$$h(x) = \langle c, x - x^0 \rangle + f(x^0)$$

satisfies $h(x) \leq f(x)$ for all $x$ and also $h(x^0) = f(x^0)$. □

## 3.6 Bumping

One of the features that distinguishes a convex point of finite order from a convex point of infinite order is stability. The next example illustrates the point.

**Example 3.24** Let

$$\Omega = \{(x, y) \in \mathbb{R}^2 : |x| < 1, |y| < 1\}.$$

Let $P$ be the boundary point $(1/2, 1)$. Let $U$ be a small open disc about $P$. Then there is no open domain $\widehat{\Omega}$ such that

(a) $\widehat{\Omega} \supseteq \Omega$ and $\widehat{\Omega} \ni P$;

(b) $\widehat{\Omega} \setminus \Omega \subseteq U$;

**(c)** $\widehat{\Omega}$ is convex.

To see this assertion, assume not. Let $x$ be a point of $\widehat{\Omega} \setminus \Omega$. Let $y$ be the point $(0.9, 0.9) \in \Omega$. Then the segment connecting $x$ with $y$ will *not* lie in $\widehat{\Omega}$.

∎

The example shows that a flat point in the boundary of a convex domain cannot be perturbed while preserving convexity. But a point of finite order *can* be perturbed:

PROPOSITION 3.25 *Let $\Omega \subseteq \mathbb{R}^N$ be a bounded, convex domain with $C^k$ boundary. Let $P \in \partial\Omega$ be a convex point of finite order $m$. Write $\Omega = \{x \in \mathbb{R}^N : \rho(x) < 0\}$. Let $\epsilon > 0$. Then there is a perturbed domain $\widehat{\Omega} = \{x \in \mathbb{R}^N : \widehat{\rho}(x) < 0\}$ with $C^k$ boundary such that*

**(a)** $\widehat{\Omega} \supseteq \Omega$;

**(b)** $\widehat{\Omega} \ni P$;

**(c)** *$\partial\widehat{\Omega} \setminus \overline{\Omega}$ consists of points of finite order $m$;*

**(d)** *The Hausdorff distance[2] between $\partial\widehat{\Omega}$ and $\partial\Omega$ is less than $\epsilon$.*

*See Figure 3.9.*

Before we begin the proof, we provide a useful technical lemma:

LEMMA 3.26 *Let $a$ be a fixed, positive number. Let $\alpha_0, \alpha_1, \ldots, \alpha_k$ and $\beta_0, \beta_1, \ldots, \beta_k$ and $\gamma_0$ be real numbers. Then there is a concave-down polynomial function $y = p(x)$ so that*

- $p(0) = \gamma_0$;
- $p(-a) = \alpha_0$, $p(a) = \beta_0$;
- $p^{(j)}(-a) = \alpha_j$ *for* $j = 1, \ldots, k$;
- $p^{(j)}(a) = \beta_j$ *for* $j = 1, \ldots, k$.

*Here the exponents in parentheses are derivatives.*

---

[2]The Hausdorff distance is a metric on the collection of compact sets. In detail, if $A$ and $B$ are compact sets, then

$$d_{\mathcal{H}}(A, B) = \max\bigl\{\sup_{a\in A}\inf_{b\in B}|a-b|, \sup_{b\in B}\inf_{a\in A}|a-b|\bigr\}.$$

See the Appendix.

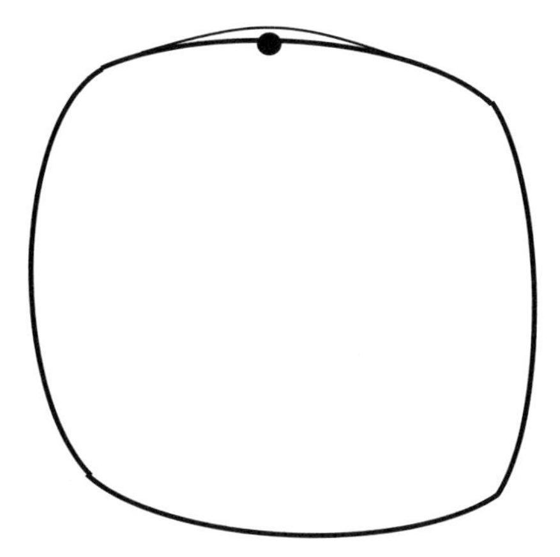

Figure 3.9: Bumping a convex point of finite order.

**Proof of Lemma 3.26:** Define

$$g_j(x) = x^{2j}$$

and let $h_j^{\theta_j}$ be the function obtained from $g_j$ by rotating the coordinates $(x, y)$ through an angle of $\theta_j$. Define

$$p(x) = c_0 - (c_1)^2 h_1^{\theta_1}(x) - (c_2)^2 h_2^{\theta_2}(x) - \cdots - (c_p)^2 h_p^{\theta_p}(x) ,$$

some positive integer $p$. If $p$ is large enough (at least $k+1$), then there will be more free parameters in the definition of $p$ than there are constants $\alpha_j$, $\beta_j$, and $\gamma_0$. So we may solve for the $c_j$ and $\theta_j$ and thereby define $p$. □

**Proof of Proposition 3.25:** First let us consider the case $N = 2$. Fix $P \in \partial\Omega$ as given in the statement of the proposition. We may assume without loss of generality that $P$ is the origin and the tangent line to $\partial\Omega$ at $P$ is the $x$-axis. We may further assume that $\Omega$ is so oriented that the boundary $\partial\Omega$ of $\Omega$ near $P$ is the graph of a concave-down function $\varphi$.

Let $\delta > 0$ be small and let $x$ and $y$ be the two boundary points that are horizontally distance $\delta$ from $P$ (situated, respectively, to the left and to the right of $P$). If $\delta$ is sufficiently small, then the angle between the tangent lines at $x$ and at $y$ will be less than $\pi/6$.

Now we think of $P = (0,0)$, $\gamma_0 = \epsilon > 0$, of $x = (-a, \alpha_0)$, and of $y = (a, \beta_0)$. Further, we set

$$\alpha_j = \varphi^{(j)}(-a) , \quad j = 1, \ldots, k$$

and

$$\beta_j = \varphi^{(j)}(a) \ , \quad j = 1, \ldots, k \, .$$

Then we may apply the lemma to obtain a concave-down polynomial $p$ which agrees with $\varphi$ to order $k$ at the points of contact $x$ and $y$.

Thus the domain $\widehat{\Omega}$ which has boundary given by $y = p(x)$ for $x \in [-a, a]$ and boundary coinciding with $\partial\Omega$ elsewhere (we are simply replacing the portion of $\partial\Omega$ which lies between $x$ and $y$ by the graph of $p$) will be a convex domain that bumps $\Omega$ *provided that* the degree of $p$ does not exceed the finite order of convexity $m$ of $\partial\Omega$ near $P$. When the degree of $p$ exceeds $m$, then the graph $y = p(x)$ may intersect $\partial\Omega$ between $x$ and $y$, and therefore not provide a geometrically valid bump.

For higher dimensions, we proceed by slicing. Let $P \in \partial\Omega$ be of finite order $m$. Let $T_P(\partial\Omega)$ be the tangent hyperplane to $\partial\Omega$ at $P$ as usual. If $\mathbf{v}$ is a unit vector in $T_P(\partial\Omega)$ and $\nu_P$ the unit outward normal vector to $\partial\Omega$ at $P$, then consider the 2-dimensional plane $\mathcal{P}_{\mathbf{v}}$ spanned by $\mathbf{v}$ and $\nu_P$. Then $\Omega_{\mathbf{v}} \equiv \mathcal{P}_{\mathbf{v}} \cap \Omega$ is a 2-dimensional convex domain which is analytically convex of order $m$ at $P \in \partial\Omega_{\mathbf{v}}$. We may apply the two-dimensional perturbation result to this domain. We do so for each unit tangent vector $\mathbf{v} \in T_P(\partial\Omega)$, noting that the construction varies smoothly with the data vector $\mathbf{v}$. The result is a smooth, perturbed domain $\widehat{\Omega}$ as desired. □

It is worth noting that the proof shows that, when we bump a piece of boundary that is analytically convex of order $m$, then we may take the bump to be analytically convex of order 2 or 4 or any degree up to and including $m$ (which of course is even).

It is fortunate that the matter of bumping may be treated more or less heuristically in the present context. In several complex variables, bumping is a more profound and considerably more complicated matter (see, for instance [BHS]).

# Exercises

**1.** Calculate the order of convexity of the boundary points of

$$\{(x_1, x_2, x_3) \in \mathbb{R}^3 : x_1^4 + x_2^4 + x_3^8 < 1\} \, .$$

**2.** What are the extreme points of the domain

$$\Omega = \{(x_1, x_2, x_3) \in \mathbb{R}^3 : x_1^4 + x_2^4 + x_3^8 < 1\} \, ?$$

**3.** What are the extreme points of the domain

$$\Omega = \{(x_1, x_2) \in \mathbb{R}^2 : x_1^2 + x_2^2 \leq 1, x_2 \geq 0\}\,?$$

**4.** Describe explicitly how to exhaust the solid unit square in the plane by smoothly bounded, strongly analytically convex domains.

**5.** TRUE or FALSE: If $f : \mathbb{R} \to \mathbb{R}$ is positive and convex then $1/f$ is never convex.

**6.** TRUE or FALSE: If $f : \mathbb{R} \to \mathbb{R}$ is positive, strictly increasing, and convex, then $f^{-1}$ is convex.

**7.** Write the function $f(x) = x^2$ explicitly as the upper envelope of affine functions.

**8.** Write out the details of Example 3.13.

**9.** It is not true that the composition of convex functions is convex. Give a counterexample.

**10.** Prove that a function $f : \mathbb{R} \to \mathbb{R}$ is convex if and only if its epigraph

$$\mathcal{E}(f) = \{(x_1, x_2) \in \mathbb{R}^2 : x_2 \geq f(x_1)\}$$

is a convex set.

**11.** Provide the details of the proof of Proposition 3.15.

# Chapter 4

# Applications of the Idea of Convexity

**Prologue:** This chapter presents some rather concrete applications of convexity theory.

Our first topic is the construction of a nowhere differentiable function. This proof is quite different from the usual treatment of the Weierstrass nowhere differentiable function (which originally arose in the context of Fourier series). It exploits convexity in a decisive but rather subtle fashion.

We then turn to the hallowed Krein-Milman theorem about extreme points. This can be thought of as a device for generating convex sets, but it is also a way to see that a good many extreme points exist. It is a result with many practical uses.

The concept of the Minkowski sum of two convex sets receives due treatment. Contrary to what we have said earlier, this is indeed an algebraic operation on our sets, and one which yields notable information for our studies. The Minkowski sum preserves convexity, and other notable properties as well. It is something that we can actually calculate explicitly in many instances.

We conclude the chapter with a treatment of the Brunn-Minkowski inequality. This is a classical result of geometric measure theory that sheds interesting light on convexity theory.

In the present chapter we present some mathematical applications of the idea of convexity. These are quite varied, and at least begin to suggest the pervasiveness and significance of the convexity concept.

## 4.1 Nowhere Differentiable Functions

Our first application (from the source [CAT]) is to the construction of nowhere differentiable functions.

THEOREM 4.1 *Let $\{a_j\}$ be a summable sequence of nonnegative real numbers. Let $\{b_j\}$ be a strictly increasing sequence of integers such that $b_j$ divides $b_{j+1}$, each $j$, and so that the sequence $\{a_j b_j\}$ does not tend to 0.*

*For each index $j \geq 1$, let $f_j$ be a continuous function mapping the real line onto the interval $[0,1]$ such that $f_j = 0$ at each even integer and $f_j = 1$ at each odd integer. For each integer $k$ and each index $j$, let $f_j$ be convex on the interval $(2k, 2k+2)$.*

*Then the continuous function*

$$f(x) \equiv \sum_{j=1}^{\infty} a_j f_j(b_j x)$$

*does not have a finite left or right derivative at any point.*

**Remark 4.2** The reader should take particular note of how convexity is used in the proof of the theorem.

**Proof of the Theorem:** Seeking a contradiction, we suppose that $f$ has a finite right derivative $f'(x)$ at the point $x$ (left derivatives are handled in the same way). Since the sequence $\{a_j b_j\}$ does not converge to zero, we can be sure that $\limsup a_j b_j > 0$. Choose $\epsilon > 0$ so that $\limsup a_j b_j > 11\epsilon$. Let $p$ be a positive number so that, if $0 < y - x < p$, then

$$\left| \frac{f(y) - f(x)}{y - x} - f'(x) \right| < \epsilon .$$

Choose an index $J > 0$ so that $a_J b_J > 11\epsilon$ and consecutive zeros of $f_J(b_j x)$ differ by less than $p/2$. This just means that $1/b_J < p/4$ (note that convexity plays a role here). Let $x_1$, $x_3$ be consecutive zeros of $f_J(b_J x)$ such that $x < x_1 < x_3$, $x_1 - x \leq x_3 - x_1 < p/2$. Let $x_2$ be the midpoint of the interval $(x_1, x_3)$. Then $x_1 - x < x_2 - x < x_3 - x < p$ and

$$x_1 - x \leq 2(x_2 - x_1) \quad , \quad x_2 - x \leq 3(x_2 - x_1) . \tag{4.1.1}$$

Let $r_1$, $r_2$, $r_3$ be real numbers defined by the equations

$$\frac{f(x_2) - f(x_1)}{x_2 - x_1} = f'(x) + r_3 ,$$

$$\frac{f(x_2) - f(x)}{x_2 - x} = f'(x) + r_2 ,$$

$$\frac{f(x_1) - f(x)}{x_1 - x_1} = f'(x) + r_1 \, .$$

Then $|r_1| < \epsilon$, $|r_2| < \epsilon$, and we have

$$\begin{aligned}
&(f'(x) + r_2)(x_2 - x) - (f'(x) + r_1)(x_1 - x) \\
&= (f(x_2) - f(x)) - (f(x_1) - f(x)) \\
&= f(x_2) - f(x_1) \\
&= (f'(x) + r_3)(x_2 - x_1) \\
&= (f'(x) + r_3)(x_2 - x) - (f'(x) + r_3)(x_1 - x) \, .
\end{aligned}$$

Hence

$$r_2(x_2 - x) - r_1(x_1 - x) = r_3(x_2 - x_1) \, .$$

From (4.1.1) we now deduce that

$$\frac{x_2 - x}{x_2 - x_1} \le 3 \quad \text{and} \quad \frac{x_1 - x}{x_2 - x_1} \le 2 \, .$$

Hence

$$|r_3| \le |r_2| \frac{x_2 - x}{x_2 - x_1} + |r_1| \frac{x_1 - x}{x_2 - x_1} \le 3\epsilon + 2\epsilon \, .$$

Therefore

$$\left| \frac{f(x_2) - f(x_1)}{x_2 - x_1} - f'(x) \right| \le 5\epsilon \, . \tag{4.1.2}$$

The same argument, with $x_3$ in place of $x_2$ (beginning again with (4.1.1)) shows that

$$\left| \frac{f(x_3) - f(x_1)}{x_3 - x_1} - f'(x) \right| \le 5\epsilon \, . \tag{4.1.3}$$

Now (4.1.2) and (4.1.3) combine to yield

$$\left| \frac{f(x_2) - f(x_1)}{x_2 - x_1} - \frac{f(x_3) - f(x_1)}{x_3 - x_1} \right| \le 10\epsilon \, . \tag{4.1.4}$$

Now fix $j < J$. Because $b_j$ divides $b_J$, we see that $x_1$ and $x_3$ lie between consecutive zeros of $f_j(b_j x)$. So $f_j(b_j x)$ is convex on the interval $(x_1, x_3)$ and so we conclude that

$$\frac{a_j(f_j(b_j x_2) - f_j(b_j x_1))}{x_2 - x_1} - \frac{a_j(f_j(b_j x_3) - f_j(b_j x_1))}{x_3 - x_1} \ge 0 \tag{4.1.5}$$

for $j < J$.

Fix now $j > J$. The points $x_1$, $x_3$ are zeros of $f_j(b_j x)$ because $b_J$ divides $b_j$. Furthermore

$$0 = f_j(b_j x_1) = f_j(b_j x_3) \le f_j(b_j x_2) \, .$$

Hence

$$\frac{a_j(f_j(b_jx_2) - f_j(b_jx_1))}{x_2 - x_1} - \frac{a_j(f_j(b_jx_3) - f_j(b_jx_1))}{x_3 - x_1} \geq 0 \qquad (4.1.6)$$

for $j > J$. By the choice of the index $J$,

$$\frac{a_J(f_J(b_Jx_2) - f_J(b_Jx_2)}{x_2 - x_1} = a_Jb_J > 11\epsilon\,.$$

Also

$$f_J(b_Jx_3) = F_J(b_Jx_1) = 0\,.$$

Therefore

$$\frac{a_J(f_J(b_Jx_2) - f_J(b_Jx_1))}{x_2 - x_1} - \frac{a_Jf_J(b_Jx_3) - f_Jb_Jx_1))}{x_3 - x_1} > 11\epsilon\,. \qquad (4.1.7)$$

We sum (4.1.5), (4.1.6), and (4.1.7) to obtain

$$\frac{f(x_2) - f(x_1)}{x_2 - x_1} - \frac{f(x_3) - f(x_1)}{x_3 - x_1} > 11\epsilon\,.$$

Now this last line is inconsistent with (4.1.4), so it follows that $f$ has no right derivative at $x$. A similar argument applies for left derivatives. □

A nice illustration of this last result is provided by $f(x) = x^2$ (thought of as a convex, increasing function from $[0,1]$ to $[0,1]$). Extend $f$ by $f(x) = f(2-x)$ for $1 < x < 2$. By iteration, this extends $f$ periodically to the entire real line. Then, by the theorem, $f(x) = \sum_j 2^{-j}f(2^jx)$ is a continuous, nowhere differentiable function.

Another interesting application is the so-called Gauss-Lucas theorem:

THEOREM 4.3 *Let $p$ be a nonconstant polynomial with complex coefficients. Then all the zeros of $p'$ (the derivative of $p$) lie in the convex hull of the zeros of $p$.*

**Remark 4.4** It is easy to see that Rolle's theorem is a special case of this result.

**Proof:** Now $p$ is a product of prime factors:

$$p(z) = \alpha\prod_{j=1}^{n}(z - a_j)\,.$$

Here $\alpha$ is the leading coefficient of the polynomial and the $a_j$ are the roots—not necessarily distinct.

At a point $z$ which is *not* a root of $p$, we may take the logarithmic derivative to write

$$\frac{p'(x)}{p(z)} = \sum_{j=1}^{n} \frac{1}{z - a_j}. \tag{4.3.1}$$

If $z$ is, in addition, a zero of $p'$, then we may write (using (4.3.1)) that

$$\sum_{j=1}^{n} \frac{1}{z - a_j} = 0 .$$

Thus

$$\sum_{j=1}^{n} \frac{\overline{z}_j - \overline{a}_j}{|z - a_j|^2} = 0 .$$

This last equation may also be written as

$$\left( \sum_{j=1}^{n} \frac{1}{|z - a_j|^2} \right) \overline{z} = \sum_{j=1}^{n} \frac{1}{|z - a_j|^2} \cdot \overline{a}_j$$

or

$$\left( \sum_{j=1}^{n} \frac{1}{|z - a_j|^2} \right) z = \sum_{j=1}^{n} \frac{1}{|z - a_j|^2} \cdot a_j .$$

Finally we may write $z$ as a weighted sum, with positive coefficients that sum to 1, of the complex numbers $a_j$;

$$z = \sum_{j=1}^{n} \frac{1/|z - a_j|^2}{\sum_{\ell=1}^{n} 1/|z - a_\ell|^2} \cdot a_j .$$

So $z$, which is a zero of $p'$, is in the convex hull of the zeros $a_j$ of $p$ itself. □

## 4.2 The Krein-Milman Theorem

The theorem discussed here is one of the most useful in convexity theory. It also has the advantage of being intuitively appealing. We introduced the result earlier in the book. Now we prove it in detail.

THEOREM 4.5 (KREIN-MILMAN) *Let $X$ be a topological vector space in which $X^*$ separates points. If $K$ is a compact, convex set in $X$, then $K$ is the closed, convex hull of its extreme points.*

**Remark 4.6** There is no harm for the reader to think of this result just for Euclidean space $\mathbb{R}^N$. But it is useful to know that it is actually true in considerable generality.

**Proof of the Theorem:** Let $\mathcal{P}$ be the collection of all compact extreme sets of $K$. Since $K \in \mathcal{P}$, we see that $\mathcal{P} \neq \emptyset$. We shall use these properties of $\mathcal{P}$ (to be proved below):

**(a)** The intersection $S$ of any nonempty subcollection of $\mathcal{P}$ is a member of $\mathcal{P}$ *unless* $S = \emptyset$.

**(b)** If $S \in \mathcal{P}$, $\Lambda \in X^*$, $\mu$ is the maximum value of $\operatorname{Re}\Lambda$ on $S$, and

$$S_\Lambda = \{x \in S : \operatorname{Re}\Lambda x = \mu\},$$

then $S_\Lambda \in \mathcal{P}$.

We observe that **(a)** is obvious.

For **(b)**, suppose that, for $x, y \in K$ and $0 < t < 1$, we have $tx + (1-t)y = z \in S_\Lambda$. Since $z \in S$ and $S \in \mathcal{P}$, we see that $x, y \in S$. Therefore $\operatorname{Re}\Lambda x \leq \mu$, $\operatorname{Re}\Lambda y \leq \mu$. Since $\operatorname{Re}\Lambda z = \mu$ and $\Lambda$ is linear, we find that $\operatorname{Re}\Lambda x = \mu = \operatorname{Re}\Lambda y$. Thus $x, y \in S_\Lambda$.

Choose some $S \in \mathcal{P}$. Let $\mathcal{P}'$ be the collection of all members of $\mathcal{P}$ that are subsets of $S$. Since $S \in \mathcal{P}'$, certainly $\mathcal{P}'$ is not empty. We partially order $\mathcal{P}'$ by set inclusion. Let $\Omega$ be a maximal totally ordered subcollection of $\mathcal{P}'$. Let $M$ be the intersection of all members of $\Omega$.

Since $\Omega$ is a collection of compact sets with the finite intersection property, $M \neq \emptyset$. By **(a)**, $M \in \mathcal{P}'$. The maximality of $\Omega$ implies that no proper subset of $M$ belongs to $\mathcal{P}$. We see from **(b)** that every $\Lambda \in X^*$ is constant on $M$. Since $X^*$ separates points of $X$, we conclude that $M$ has only one point. Thus $M$ is an extreme point of $K$. □

**Example 4.7** A closed cube in $\mathbb{R}^N$ is the closed, convex hull of its vertices.

A closed ball in $\mathbb{R}^N$ is the closed, convex hull of *all* its boundary points (that is to say, it is the closed, convex hull of the sphere). ∎

Certainly the Krein-Milman theorem guarantees that any compact, convex set has plenty of extreme points. That in itself is valuable information.

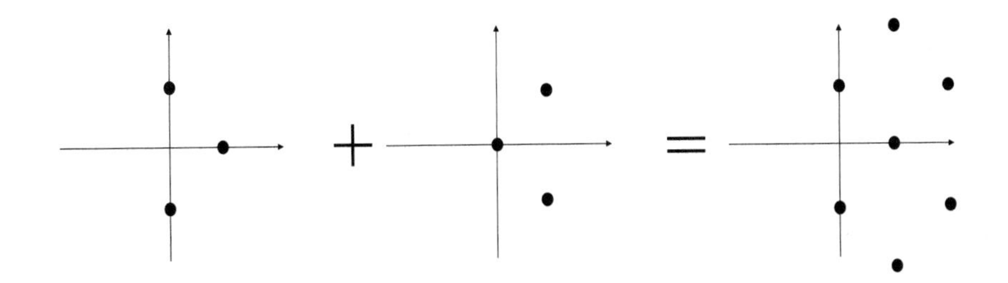

Figure 4.1: A Minkowski sum.

## 4.3 The Minkowski Sum

We shall see a notable application of the idea of Minkowski sum in the Brunn-Minkowski inequality in the next section. But in fact this idea has many interesting applications, including those in robotics and computer graphics. We shall indicate some of these at the end of the section.

**Definition 4.8** Let $A$ and $B$ be sets in $\mathbb{R}^N$. Then the *Minkowski sum* of $A$ and $B$ is the set

$$A + B = \{a + b : a \in A, b \in B\}.$$

**Example 4.9** Let

$$A = \{(1,0),(0,1),(0,-1)\} \subseteq \mathbb{R}^2$$

and

$$B = \{(0,0),(1,1),(1,-1)\} \subseteq \mathbb{R}^2.$$

Then the Minkowski sum is

$$\begin{aligned} A + B &= \{(1,0),(2,1),(2,-1),(0,1),(1,2), \\ &\qquad (1,0),(0,-1),(1,0),(1,-2)\}. \end{aligned}$$

See Figure 4.1 for the idea. There are in principle nine points in the sum, but the point $(1,0)$ is repeated. And $(1,0)$ is *not* a vertex of the hexagon.

Notice that each of $A$ and $B$ consists of the vertices of a triangle, while $A + B$ consists of the vertices of a hexagon. ∎

Now we present a few intuitively appealing properties of the Minkowski sum.

PROPOSITION 4.10 *If $A$ and $B$ are convex then their Minkowski sum is also convex.*

**Proof:** Let $x \in A + B$ and $y \in A + B$. Then $x = a + b$, for some $a \in A$ and $b \in B$, and $y = a' + b'$, for some $a' \in A$ and $b' \in B$. So the segment $\varphi(t) = (1 - t)a + ta'$, $0 \leq t \leq 1$ lies in $A$ and the segment $\widetilde{\varphi}(t) = (1 - t)b + tb'$ lies in $B$.

It follows then that the segment $\psi(t) = (1 - t)(a + b) + t(a' + b')$ lies in $A + B$. This is of course the segment that connects $x$ to $y$. So we have proved from first principles that $A + B$ is geometrically convex according to our classical definition. □

PROPOSITION 4.11 *If $A$ and $B$ are path connected then so is $A + B$.*

**Proof:** The proof is just the same as that for the preceding proposition, with linear paths replaced by arbitrary paths. □

PROPOSITION 4.12 *Let $A$ and $B$ be sets in $\mathbb{R}^N$. Then the convex hull of $A + B$ is the Minkowski sum of the convex hulls of $A$ and $B$.*

**Proof:** We treat the special case when $A$ and $B$ are finite sets. The general case then follows by a limiting argument (i.e., exhaust $A$ and $B$ by finite nets).

Using the Krein-Milman theorem (see Section 4.1 above), we may replace each of the finite collections $A$ and $B$ with the collection of the extreme points. Their theorem says that the convex hull of the full set is the same as the convex hull of the extreme points.

Notice that if $x$, $x'$ are extreme points of $A$ and $y$, $y'$ are extreme points of $B$, then we may consider the Minkowski sum of $\overline{xx'}$ and $\overline{yy'}$. Of course these are, respectively, the segments determined by $x, x'$ and $y, y'$. The result of the sum is a parallelogram.

Since $A$ and $B$ are convex (in fact convex polygons), we can be sure that $A + B$ is convex by Proposition 4.9 above. By the remarks in the last paragraph, we in fact may conclude that $A + B$ is a convex polygon. In particular, $A + B$ is equal to its own convex hull. That proves our result. □

The Minkowski sum plays a central role in mathematical morphology. It arises in the brush-and-stroke paradigm of 2D computer graphics (with various uses, notably by Donald E. Knuth in Metafont), and as the solid sweep operation of 3D computer graphics.

Motion planning Minkowski sums are used in motion planning of an object among obstacles. They are used for the computation of the configuration space, which is the set of all admissible positions of the object. In the simple model of translational motion of an object in the plane, where the position of an object may be uniquely specified by the position of a fixed point of this object, the configuration space is the Minkowski sum of the set of obstacles and the movable object placed at the origin and rotated 180 degrees.

In numerical control (NC) machining, the programming of the NC tool exploits the fact that the Minkowski sum of the cutting piece with its trajectory gives the shape of the cut in the material.

## 4.4 The Brunn-Minkowski Inequality

The result that we now consider is a fundamental inequality from geometric measure theory that is proved using very basic convexity (or concavity) considerations.

In what follows, if $A$ and $B$ are sets in $\mathbb{R}^N$, we set

$$A + B = \{a + b : a \in A, b \in B\} .$$

We let $|A|$ denote the Lebesgue outer measure[1] of $A$.

THEOREM 4.13 (BRUNN-MINKOWSKI) *If $A$ and $B$ are nonempty Lebesgue measurable subsets of $\mathbb{R}^N$, then*

$$|A + B|^{1/N} \geq |A|^{1/N} + |B|^{1/N} .$$

**Proof:** Let $\mathcal{F}$ denote the family of all rectangular boxes

$$P_1 \times P_2 \times \cdots \times P_n ,$$

where each $P_j$ is a nonempty, bounded, open subinterval in $\mathbb{R}$.

If $A = P_1 \times P_2 \times \cdots \times P_N$ and $B = Q_1 \times Q_2 \times \cdots \times Q_N$ are both elements of $\mathcal{F}$, then notice that

$$A + B = (P_1 + Q_1) \times (P_2 + Q_2) \times \cdots \times (P_N + Q_N) .$$

Now set $u_j = |P_j|/|P_j + Q_j|$ and $v_j = |Q_j|/|P_j + Q_j, j = 1, \ldots, N$. We use the standard fact that the geometric mean is majorized by the arithmetic mean

[1]The reader unfamiliar with measure theory can just think of the area or volume of the set.

(see [FED, § 2.4.13]) to see that

$$\begin{aligned}
&\left[|A|^{1/N} + |B|^{1/N}\right] \cdot |A+B|^{-1/N} \\
&= \prod_{j=1}^{N} u_j^{1/N} + \prod_{j=1}^{N} v_j^{1/N} \\
&\le \sum_{j=1}^{N} u_j/N + \sum_{j=1}^{N} v_j/N \\
&= 1\,.
\end{aligned}$$

Here we have used the fact that $u_j + v_j = 1$.

Next we turn to the special case of our result when $A$ is the finite disjoint union of a family $\mathcal{G}$ in $\mathcal{F}$ and $B$ is the finite disjoint union of a family $\mathcal{H}$ in $\mathcal{F}$. We shall apply induction with respect to $\text{card}(\mathcal{G}) + \text{card}(\mathcal{H})$.

If $\text{card}(\mathcal{G}) > 1$ then we choose $j \in \{1, 2, \ldots, N\}$ and $a \in \mathbb{R}$ so that each of the two sets

$$A_1 \equiv A \cap \{x \in \mathbb{R} : x_j < a\}$$

and

$$A_2 \equiv A \cap \{x \in \mathbb{R} : x_j > a\}$$

contains some member of }. We also choose $b \in \mathbb{R}$ so that the sets

$$B_1 \equiv B \cap \{x \in \mathbb{R} : x_j < b\}$$

and

$$B_2 \equiv B \cap \{x \in \mathbb{R} : x_j > b\}$$

satisfy the equations

$$|A_k|/|A| = |B_k|/|B| \quad \text{for} \quad k = 1, 2\,.$$

Now define

$$\mathcal{G}_k = \{U \cap A_k : U \in \mathcal{G}, U \cap A_k \neq \emptyset\}$$

and

$$\mathcal{H}_k = \{V \cap B_k : V \in \mathcal{H}, V \cap B_k \neq \emptyset\}\,.$$

We see that $A_k = \cup \mathcal{G}_k$ and $B_k = \cup \mathcal{H}_k$; also

$$\text{card}(\mathcal{G}_k) < \text{card}(\mathcal{G}) \quad \text{and} \quad \text{card}(\mathcal{H}_k) \le \text{card}(\mathcal{H})\,.$$

Since $A_1 + B_1$ and $A_2 + B_2$ are separated by $\{x : x_j = a + b\}$, induction yields that

$$\begin{aligned} |A + B| &\geq |A_1 + B_1| + |A_2 + B_2| \\ &\geq \left[|A_1|^{1/N} + |B_1|^{1/N}\right]^N \\ &\quad + \left[|A_2|^{1/N} + |B_2|^{1/N}\right]^N \\ &= \left[|A|^{1/N} + |B|^{1/N}\right]^N . \end{aligned}$$

Now standard outer regularity results from measure theory allow us to pass from the simple case of sets that are finite disjoint unions of elements of $\mathcal{G}$ and $\mathcal{F}$ to the case when $A$ and $B$ are arbitrary compact sets. And then we may again invoke outer regularity to pass to the case when $A$ and $B$ are Lebesgue measurable. □

Even a simple example illustrates Brunn-Minkowski nicely, and shows that it is sharp.

**Example 4.14** For $a > 0$, et $A \subseteq \mathbb{R}^N$ be the set $\{(x_1, x_2, \ldots, x_N) : 0 \leq x_j \leq a, j = 1, 2, \ldots, N\}$ and let $B \subseteq \mathbb{R}^N$ be the set $\{(x_1, x_2, \ldots, x_N) : 0 \leq x_j \leq a, j = 1, 2, \ldots, N\}$. Then

$$|A + B|^{1/N} = 2a ,$$

$$|A|^{1/N} = a ,$$

and

$$|B|^{1/N} = a .$$

So the Brunn-Minkowski inequality says in this case that $2a \geq a + a$. See Figure 4.2. ∎

# Exercises

1. Calculate the Minkowski sum of a square with interior and a closed disc.
2. Calculate the Minkowski sum of the $x$-axis and the $y$-axis.
3. Calculate both sides of the Brunn-Minkowski inequality when $A$ and $B$ are discs. Does it matter whether or not the discs are disjoint?

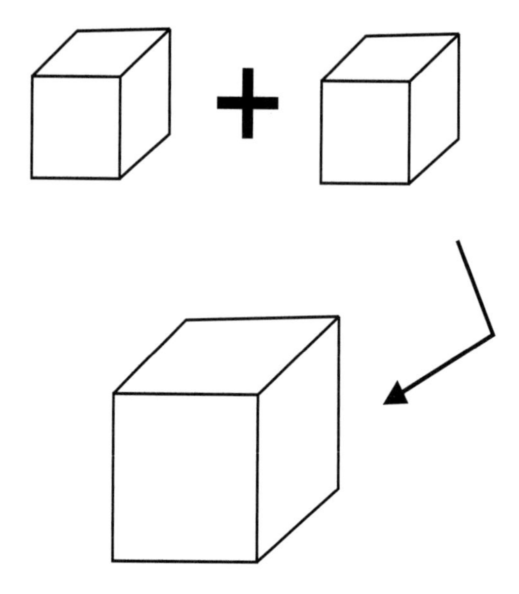

Figure 4.2: The Brunn-Minkowski inequality.

**4.** If $S$ is a convex polygon with interior in the plane then $S$ is the closed, convex hull of its extreme points. What are those extreme points?

**5.** Are there circumstances in the Krein-Milman theorem where the domain is the closed, convex hull of *some* of the extreme points but *not all* of the extreme points?

**6.** Give an example of a closed, convex set with just two extreme points, and so that the set is the closed, convex hull of those two extreme points.

**7.** The mapping

$$(x_1, x_2) \longmapsto (x_1^2 + x_2^2, x_1^2 - x_2^2)$$

does *not* preserve compact sets. Give an example.

**8.** Give an example of a compact, convex set in the plane with countably many extreme points, and so that the domain is the closed convex hull of those extreme points.

**9.** Suppose that $\Omega$ is a compact, convex set in the plane with smooth boundary. Let $R > 0$. Assume that each boundary point $P$ has the property that there is an internally tangent disc to the boundary, tangent at $P$, of radius $R$. Show that $\Omega$ is the Minkowski sum of the disc $\overline{D}(0, R)$ and some other compact, convex set.

**10.** Give an example for which the Brunn-Minkowski inequality is a strict inequality.

# Chapter 5

# More Sophisticated Ideas

**Prologue:** In our penultimate chapter we treat a number of advanced topics in convexity theory.

We begin with the concept of the polar of a set and indicate some of its uses—in particular with regard to gauges or distance functions.

We give a fairly thorough introduction to George Dantzig's simplex method for studying linear programming problems. Often termed one of the most important algorithms of modern times, the simplex method uses convexity in decisive ways to analyze the scheduling of airlines, the design of routing, and for many other very practical problems.

We treat various generalizations of convex sets, such as star-like sets and ortho-convex sets. These show the flexibility of the collection of ideas that we have been studying.

We treat an integral formula for convex functions, and also make some studies of the classical Gamma function. This and some other hard analytic facts round out our studies and, except for one special topic in the terminal Chapter 6, draw this book to a propitious close.

## 5.1 The Polar of a Set

Let $S \subseteq \mathbb{R}^N$ be any set. Then the *polar* of $S$ is defined to be

$$S^o = \{\alpha \in \mathbb{R}^N : \alpha \cdot x \leq 1 \text{ for all } x \in S\} .$$

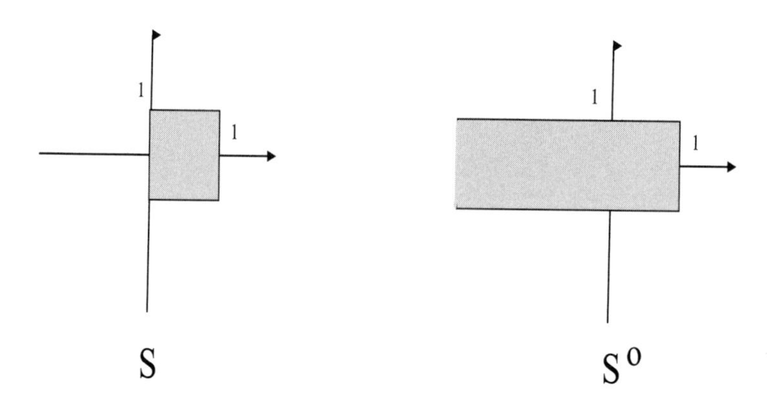


Figure 5.1: The polar of a set.

See Figure 5.1.

The first interesting fact about the polar is as follows:

PROPOSITION 5.1 *Let $S \subseteq \mathbb{R}^N$ be any set. Then the polar $S^o$ of $S$ is convex. Also the polar contains the origin.*

**Proof:** The second assertion is obvious.

For the first, let $\alpha_1, \alpha_2 \in S^o$. Then, for $0 \le t \le 1$ and $x \in S$,

$$\begin{aligned} \big[(1-t)\alpha_1 + t\alpha_2\big] \cdot x &= (1-t)\big[\alpha_1 \cdot x\big] + t\big[\alpha_2 \cdot x\big] \\ &\le (1-t) \cdot 1 + t \cdot 1 = 1 \,. \end{aligned}$$

Hence $(1-t)\alpha_1 + t\alpha_2 \in S^o$ and $S^o$ is convex. □

**Example 5.2** Let $v_1, v_2, \ldots, v_k$ be finitely many points in $\mathbb{R}^N$. Then the polar of the convex hull of $\{v_1, v_2, \ldots, v_k\}$ is the polyhedron $\{\alpha \in \mathbb{R}^N : \alpha \cdot v_1 \le 1, \alpha \cdot v_2 \le 1, \ldots, \alpha \cdot v_k \le 1\}$.

In fact this is nearly obvious. Because an element of the convex hull of $\{v_1, v_2, \ldots, v_k\}$ is $s_1 v_1 + s_2 v_2 + \cdots s_k v_k$ for $s_j$ nonnegative and summing to 1. And thus an element $\alpha$ of the polar of that set satisfies

$$s_1 \alpha \cdot v_1 + s_2 \alpha \cdot v_2 + \cdots + s_k \alpha \cdot v_k \le 1 \,.$$

And this is equivalent to our desired conclusion. ∎

An ancillary but useful idea is the following.

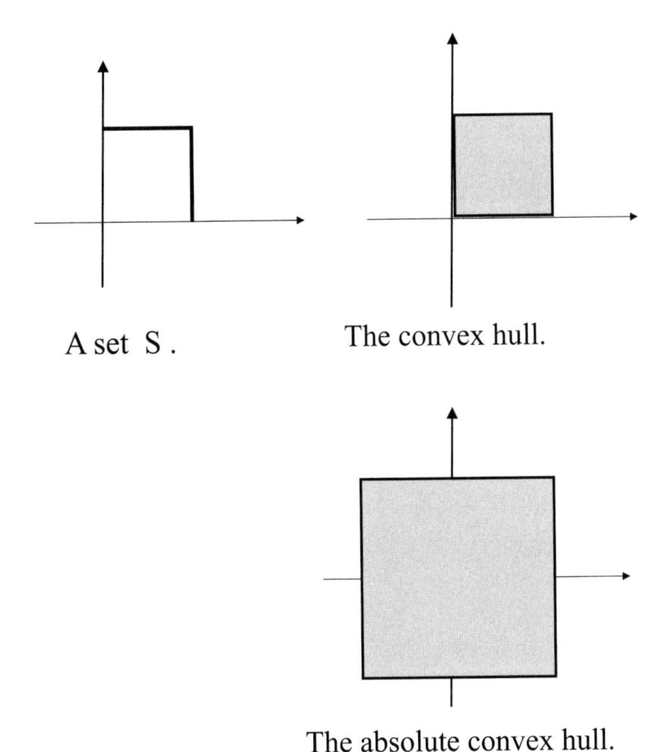


Figure 5.2: The absolute convex hull vs. the convex hull.

**Definition 5.3** A set $S \subseteq \mathbb{R}^N$ is *absolutely convex* if, whenever $x_1, x_2 \in S$ and $|\lambda_1| + |\lambda_2| \leq 1$, then $\lambda_1 x_1 + \lambda_2 x_2 \in S$.

The *absolute convex hull* of a set is the intersection of all absolutely convex sets that contain it. See Figure 5.2.

**Example 5.4** Let

$$S = \{(t_1, t_2) \in \mathbb{R}^2 : 0 \leq t_1 \leq 1, 0 \leq t_2 \leq 1\} .$$

Then $S$ is a square with boundary, so it is obviously convex. But it is *not* absolutely convex. For let $x_1 = (1/2, 0)$ and $x_2 = (1/2, 0)$. Also let $\lambda_1 = \lambda_2 = -1/2$. Then

$$\lambda_1 x_1 + \lambda_2 x_2 = (-1/2, 0) \notin S .$$

∎

And now some further properties of the polar:

PROPOSITION 5.5 *Let $A$, $B$, $A_j$ be any sets in $\mathbb{R}^N$. Then*

**(a)** *If $A \subseteq B$, then $B^o \subseteq A^o$;*

**(b)** *If $\delta > 0$ and $\delta A$ represents the dilation of $A$ by a factor of $\delta$, then $(\delta A)^o = (1/\delta)A^o$;*

**(c)** $(\cup_j A_j)^o = \cap_j A_j^o$;

**(d)** *The set $A^o$ is absolutely convex;*

**(e)** *The bipolar of $A$, denoted by $A^{oo}$, is the absolutely convex hull of $A$.*

**Proof:** We verify the assertions in order:

**Proof of (a):** Assume that $A \subseteq B$ and let $\alpha \in B^o$. So $\alpha \cdot x \leq 1$ for all $x \in B$. So certainly $\alpha \cdot x \leq 1$ for all $x \in A$. Hence $\alpha \in A^o$.

**Proof of (b):** Let $\alpha \in (\delta A)^o$. Then $\alpha \cdot (\delta x) \leq 1$ for all $x \in A$. Hence $(\delta\alpha) \cdot x \leq 1$ for all $x \in A$. In other words, $(\delta\alpha) \in A^o$ or $\alpha \in (1/\delta)A^o$.

**Proof of (c):** If $\alpha \in (\cup_j A_j)^o$, then $\alpha \cdot x \leq 1$ for $x$ in any of the $A_j$. Hence

$$\forall j, \alpha \cdot x \leq 1 \text{ for } x \in A_j \,.$$

Thus

$$\alpha \in \cap_j A_j^o \,.$$

The proof of the converse direction is similar.

**Proof of (d):** Let $p_1, p_2 \in A^o$. Consider constants $\lambda_1, \lambda_2$ with $|\lambda_1|+|\lambda_2| \leq 1$. Since $p_1 \cdot x \leq 1$ for all $x \in A$ and $p_2 \cdot x \leq 1$ for all $x \in A$, we see that

$$\lambda_1 p_1 \cdot x + \lambda_2 p_2 x \leq 1$$

for all $x \in A$. Hence

$$(\lambda_1 p_1 + \lambda_2 p_2) \cdot x \leq 1$$

for all $x \in A$. So $\lambda_1 p_1 + \lambda_2 p_2 \in A^o$.

**Proof of (e):** We leave this argument for the interested reader. □

Figure 5.3: A cone.

**Example 5.6** Let $S \subseteq \mathbb{R}^N$ be the singleton $\{x_0\}$. Then it is clear that the polar of $S$ is the half space

$$S^o = \{x \in \mathbb{R}^N : x \cdot x_0 \leq 1\}.$$

If, for simplicity, we take $x_0 = (1, 0, 0, \ldots, 0)$, then it is clear that

$$S^o = \{(x_1, x_2, \ldots, x_N) : x_1 \leq 1\}$$

and

$$S^{oo} = \{(x, 0, 0, \ldots, 0) : x \leq 1\}.$$

Notice that $S^{oo}$ is the absolute convex hull of $S$. ∎

We say that a set $A \subseteq \mathbb{R}^N$ is a *cone* if, whenever $x \in A$ and $\lambda \geq 0$, then $\lambda x \in A$. See Figure 5.3.

PROPOSITION 5.7 *If $A \subseteq \mathbb{R}^N$ is a cone then $A^o$ is also a cone.*

**Proof:** Let $\alpha \in A^o$. Then, by definition,

$$\alpha \cdot x \leq 1$$

for all $x \in A$. Now, if $\lambda > 0$ (we handle the case $\lambda = 0$ separately) then we see that

$$(\lambda\alpha) \cdot x \leq \lambda$$

for all $x \in A$. But we may rewrite this as

$$(\lambda\alpha) \cdot (x/\lambda) \leq 1$$

for all $x \in A$. Since $A$ is assumed to be a cone, we see that $t \equiv x/\lambda \in A$. So we may rewrite the last line as

$$(\lambda\alpha) \cdot t \leq 1$$

for all $t \in A$. That proves the result for $\lambda \neq 0$. But the case $\lambda = 0$ is trivial, so we are done. □

In Sections 1.3 and 2.2 we discussed the Minkowski functional. Right now we shall briefly treat the concept of gauge, which is closely related.

**Definition 5.8** If $K$ is a convex set, then define

$$\|x\| = \inf\{\lambda \in (0, \infty) : \lambda^{-1}x \in K\}.$$

We call $\| \quad \|$ a *gauge*.

Clearly a gauge is a seminorm.[1] Also we have

PROPOSITION 5.9 *We have the relations*

$$\{x : \|x\| \leq 1\} = \bigcap_{\lambda > 1} \lambda K$$

*and*

$$\{x : \|x\| < 1\} = \bigcup_{\lambda < 1} \lambda K.$$

**Proof:** We prove the first of these. Now, if $\mu > 0$, then

$$\begin{aligned} \|\mu x\| &= \inf\{\lambda : \lambda^{-1}\mu x \in K\} \\ &= \inf\{\lambda\mu : (\lambda\mu)^{-1}\mu x \in K\} \\ &= \inf\{\lambda\mu : \lambda^{-1}x \in K\} \\ &= \mu\|x\|. \end{aligned}$$

Thus $\| \quad \|$ is homogeneous of degree 1. Also $\|x\| = \|-x\|$. So we see that

$$\begin{aligned} \|x : \|x\| \leq 1\} &= \{x : \inf\{\lambda : \lambda^{-1}x \in K\} \leq 1\} \\ &= \{x : \lambda^{-1}x \in K \text{ for all } \lambda > 1\} \\ &= \bigcap_{\lambda > 1} \lambda K. \end{aligned}$$

[1] A seminorm is like a norm, but it is not assumed to satisfy the property $\|x\| = 0$ implies $x = 0$. See the Appendix.

It is not difficult to see that $\| \ \|$ is a seminorm.

We leave the proof of the other assertion to the reader. □

**Example 5.10** Let $E_1 = \{(x_1, x_2) \in \mathbb{R}^2 : x_2^2 + x_2^2 < 1\}$ and $E_2 = \{(x_1, x_2) \in \mathbb{R}^2 : x_2^2 + x_2^4 < 1\}$. We wish to construct two different metrics on $\mathbb{R}^2$ modeled on these domains (thought of as unit balls).

We define

$$d_1((s_1, s_2), (t_1, t_2)) = \sqrt{(s_1 - t_1)^2 + (s_2 - t_2)^2}$$

and

$$d_2((s_1, s_2), (t_1, t_2)) = \sqrt[4]{(s_1 - t_1)^2 + (s_2 - t_2)^4}\,.$$

Then the space $X_1$ given by $\mathbb{R}^2$ equipped with the metric $d_1$ is comparable to the gauge modeled on the ball $E_1$. Likewise, the space $X_2$ given by $\mathbb{R}^2$ equipped with the metric $d_2$ is comparable to the gauge modeled on the ball $E_2$.

It is a matter of some interest to see that $X_1$ and $X_2$ are not linearly equivalent. But this is obvious because $E_1$ is a convex set of order 2 and $E_2$ is convex set of order 4 (see Section 3.2). If the two spaces were linearly equivalent then their unit balls would have to be comparable, and they are not. See Figure 5.4. ∎

The discussion in the last example is of more than philosophical interest. For, in the function theory of several complex variables, one considers the geometry coming from the singularity of the Poisson kernel (this is an isotropic geometry comparable to that coming from $E_1$ above) and also the geometry coming from the singularity of the Bergman kernel (this is a nonisotropic geometry comparable to that coming from $E_2$ above). That these two geometries are naturally inequivalent has profound consequences for complex analysis.

It is a matter of some interest to see the relation between gauges and polars.

THEOREM 5.11 *Let $S \subseteq \mathbb{R}^N$ be a closed, convex set so that $x \in S$ implies that $-x \in S$ (we call such a set balanced). Let $\| \ \|$ be the gauge induced by $S$. Then*

$$\|x\| = \sup_{y \in S^o} |y \cdot x|\,. \tag{5.11.1}$$

Thus we see that the polar of a set acts like the unit ball of a seminorm. Of course such unit balls are automatically convex.

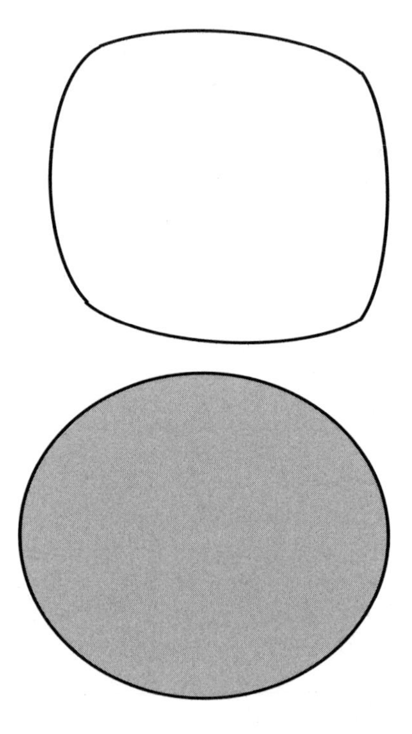

Figure 5.4: Inequivalent unit balls.

**Proof of the Theorem:** Since $S$ is closed, we see that $\mathbb{R}^N \setminus S$ is open. Of course

$$y \mapsto y \cdot x$$

is a continuous linear functional. It follows then that $S^{oo} = S$.

Let $\| \ \|_S$ be the right-hand side of (5.11.1). Then this function is clearly convex and homogeneous of degree 1. Since $S = \{x : \|x\| \leq 1\}$, we need only show that $S = \{x : \|x\|_S \leq 1\}$. But we note that $\|x\|_S \leq 1$ if and only if $x \in S^{oo} = S$. That ends the argument. □

**Example 5.12** It is perhaps natural to wonder for which sets $S$ it holds that $S = S^o$. There are many examples of such sets.

Consider for example

$$S = \{x \in \mathbb{R}^N : x_1 \geq 0\} .$$

Then it is straightforward to check that $S = S^o$. Moreover, is $\lambda$ is any orthogonal transformation then $\lambda S$ has the property that $\lambda S = (\lambda S)^o$. One can also verify directly that the closed unit ball $B = \{x \in \mathbb{R}^N : \|x\| \leq 1\}$ is self-polar.

For which $S$ does it hold that $S^o = -S$? It can be shown with a tedious calculation that there is only one such set, and it is again the closed unit ball $\overline{B}$. ▮

# 5.2 Optimization

## 5.2.1 Introductory Thoughts

One part of mathematics that relies decisively on convexity theory is optimization. Optimization is a *very* applied part of mathematics. It is used to schedule airlines, manage supply lines, and govern other very practical activities.

Although optimization theory is now an entire subject area of mathematics unto itself, one of the key parts of the discipline, and one of the pioneering parts, is linear programming. The name usually associated with linear programming is George Dantzig, although some notable preliminary work was done by Leonid Kantorovich. In fact Kantorovich's work was kept confidential because of security considerations in World War II. It was Dantzig who first published about linear programming in 1947, and he continued to develop the theory of linear programming for the next fifty years or more. We

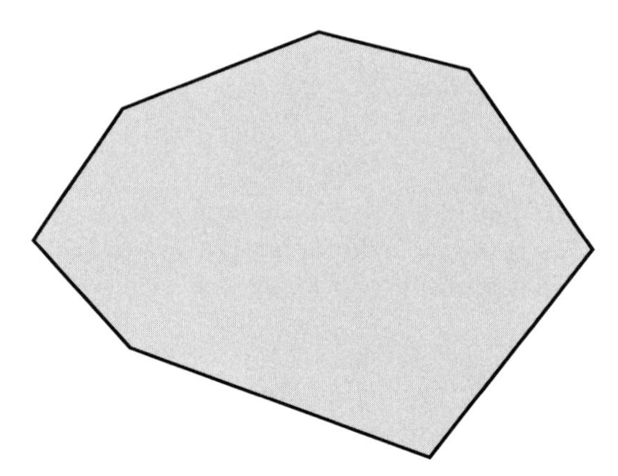

Figure 5.5: The constraints for a linear programming problem in dimension two.

shall describe the simplex method here as an illustration of the importance of convexity in optimization.

It is easy to see that an important property of convex functions is that any extremum must be a minimum, and any local minimum must be a global minimum. Both these observations play a key role in the discussion that follows.

Generally speaking we wish to consider the problem of maximizing (or minimizing) a linear function $f$ over a region $\mathcal{U}$ in Euclidean space which is defined as the intersection of finitely many half-spaces. Refer to Figure 5.5 for the geometry of the situation in $\mathbb{R}^2$. Obviously such a region is convex. It is also clear, after a moment's thought, that the extrema of $f$ will occur at the vertices of the boundary of $\mathcal{U}$.

What is interesting here is that, in a typical application, there will be tens of thousands or even hundreds of thousands of vertices. So it is not generally practical to just check each vertex to find the maximum or minimum that we seek. A more sophisticated idea is required.

A favorite example, and in fact one that Dantzig gave in his original paper, was this: You have 70 people and 70 jobs and you want to find the best assignment of each of those 70 people to a job. Of course the number of different possible assignments is 70!, and that number is in fact greater than the number of particles in the observable universe. It would take a fast computer quite a long time to check all these possiblities. But the problem can be solved with linear programming (the simplex method) in just a few seconds. The theory of linear programming drastically reduces the number of cases that must be checked. It is a truly profound idea.

### 5.2.2 Setup for the Simplex Method

It is important to understand that the function to be extremized is a scalar-valued linear function on $\mathbb{R}^N$. So it has the form

$$f(x_1, x_2, \ldots, x_N) = c_1 x_1 + c_2 x_2 + \cdots + c_N x_N .$$

It is more interesting to think about how to describe the constraint region analytically.

The simplest fashion in which to describe the $k$ half-spaces whose intersection forms the constraint space is

$$\begin{array}{ccccccc} a_{11}x_1 & + & a_{12}x_2 & + & \cdots & + & a_{1N}x_N \leq b_1 \\ a_{21}x_1 & + & a_{22}x_2 & + & \cdots & + & a_{2N}x_N \leq b_2 \\ a_{31}x_1 & + & a_{32}x_2 & + & \cdots & + & a_{3N}x_N \leq b_3 \\ & & \cdots & & & & \\ a_{k1}x_1 & + & a_{k2}x_2 & + & \cdots & + & a_{kN}x_N \leq b_k . \end{array}$$

Finally it is standard—and this hypothesis makes good sense in the context of the intended applications—to assume that $x_1 \geq 0$, $x_2 \geq 0$, $\ldots$, $x_N \geq 0$.

In summary, the setup for our linear programming problem (written now in standard matrix notation) is

$$\max\left\{ {}^t C x : Ax \leq b, x \geq 0 \right\} .$$

We next provide a very standard example, just to illustrate how a linear programming problem arises in practice.

**Example 5.13** A farmer has a piece of land which is $L$ km$^2$ in area. It is to be planted with corn or flax or perhaps some combination of the two. We want to maximize the farmer's profit.

The constraints are these. First of all, the farmer has only $F$ kilograms of fertilizer and $P$ kilograms of pesticide. Every square kilometer of corn requires $F_1$ kilograms of fertilizer and $P_1$ kilograms of insecticide. Every square kilometer of flax requires $F_2$ kilograms of fertilizer and $P_2$ kilograms of pesticide. We let $S_1$ be the selling price of corn per square kilometer and $S_2$ the selling price of flax per square kilometer.

Finally we let $x_1$ be the number of square kilometers planted with corn and $x_2$ the number of square kilometers planted with flax. The important point to understand now is that we can maximize the profit by choosing $x_1$ and $x_2$ appropriately.

We now express our problem as a linear programming problem:

**to be maximized:** $S_1 \cdot x_1 + S_2 \cdot x_2$

**constraints:** $x_1 + x_2 \leq L$
$F_1 \cdot x_1 + F_2 \cdot x_2 \leq F$
$P_1 \cdot x_1 + P_2 \cdot x_2 \leq P$
$x_1 \geq 0\ ,\ \ x_2 \geq 0\,.$

We leave it as an exercise for you to write these conditions in matrix form. ∎

### 5.2.3 Augmented Form

Before a linear programming problem can be attacked with the simplex method, it must be converted to *augmented form*. This is a technique of introducing additional variables (called *slack variables*) so as to convert the inequalities (in the constraint) to equalities. We illustrate the idea by converting the example of the farmer in Example 5.13 to augmented form:

**Example 5.14** In Example 5.13, we introduce new nonnegative variables $x_3$, $x_4$, $x_5$. The problem then becomes

**to be maximized:** $S_1 \cdot x_1 + S_2 \cdot x_2$

**constraints:** $x_1 + x_2 + x_3 = L$
$F_1 \cdot x_1 + F_2 \cdot x_2 + x_4 = F$
$P_1 \cdot x_1 + P_2 \cdot x_2 + x_5 = P$
$x_1 \geq 0\ ,\ \ x_2 \geq 0\ x_3 \geq 0\ ,\ \ x_4 \geq 0\ ,\ \ x_5 \geq 0\,.$

We see that an inequality becomes an equality just by mandating that the difference of the greater side and the lesser side be nonnegative.

In matrix form, the linear programming problem now becomes to maximize

$Z$ for the equation

$$\begin{bmatrix} 1 & -S_1 & -S_2 & 0 & 0 & 0 \\ 0 & 1 & 1 & 1 & 0 & 0 \\ 0 & F_1 & F_2 & 0 & 1 & 0 \\ 0 & P_1 & P_1 & 0 & 0 & 1 \end{bmatrix} \begin{bmatrix} Z \\ x_1 \\ x_2 \\ x_3 \\ x_4 \\ x_5 \end{bmatrix} = \begin{bmatrix} 0 \\ L \\ F \\ P \end{bmatrix} \cdot \begin{bmatrix} x_1 \\ x_2 \\ x_3 \\ x_4 \\ x_5 \end{bmatrix} \geq 0 , \qquad (5.14.1)$$

where $x_1 \geq 0$, $x_2 \geq 0$, $x_3 \geq 0$, $x_4 \geq 0$, $x_5 \geq 0$. ∎

### 5.2.4 The Simplex Algorithm

The simplex algorithm has been called one of the ten most important algorithms of the twentieth century. It is significant, but it is quite simple. It is based on Gaussian elimination. That is, it is based on operations that you can perform on a system of linear equations that will preserve the solution set. These operations include **(i)** multiplying one equation by a constant and **(ii)** adding to one equation a constant multiple of another.

Now suppose that we are given an optimization problem. That is, a linear programming problem as described at the end of the last subsection. The first step in our solution process is to turn the system of equations into what is called a *feasible array*. So we introduce the slack variables as usual to turn the system of inequalities that describes the constraints into equalities. Then we turn this system of equations into an *array*. Well, the array is simply the matrix on the left-hand side of (5.14.1).

In fact we shall customarily augment this matrix by adding a column on the right, with a vertical bar setting it off. The rightmost column will be our data. Thus our feasible array is

$$\left[\begin{array}{cccccc|c} 1 & -S_1 & -S_2 & 0 & 0 & 0 & 0 \\ 0 & 1 & 1 & 1 & 0 & 0 & L \\ 0 & F_1 & F_2 & 0 & 1 & 0 & F \\ 0 & P_1 & P_1 & 0 & 0 & 1 & P \end{array}\right] \qquad (5.15)$$

Notice that all but one of the coefficients in the first row of this array are negative. That is because this row comes from the formula for the function that we are trying to optimize. We will always have negative coefficients in the first row when we begin an optimization problem of this sort. The goal of the iterative pivoting process is to eliminate those negative coefficients. When they are all gone—that is to say, when all the entries in the first row are nonnegative—then the pivoting process is complete and we can read off the solution of the problem.

And now we very briefly describe the pivoting algorithm. A good source, with very detailed examples, is [AHR].

**(a)** First we select the column with the most negative indicator in the first row. Referring to display (5.15), we see that this could be column 2 or column 3 in the example of the farmer. Let us select column 2. We call this the *pivot column.*

**(b)** Now we divide each row by the corresponding element in the pivot column.

**(c)** The pivot element is, by definition, the intersection of the pivot column and the row with the least number to the right of the vertical line.

**(d)** Divide the row containing the pivot element (the pivot row) by the pivot element.

**(e)** Subtract a suitable multiple of the pivot row from each of the other rows so that all the lead (i.e., leftmost) elements are 0.

Now you will find that the most negative element in the first row is *not as negative* as it was in the first step. So we have effected an improvement. A first-order solution to our optimization problem is the number that appears in the first row to the right of the vertical bar.

If there are in fact no negative elements left in the first row, then our solution process is finished and the first-order solution found in the last paragraph is *the* solution of the optimization problem. If there remain some negative elements in the first row, then select the most negative and repeat the pivoting process described above.

Keep going until there are no negative elements left in the first row. Then the number to the right of the bar in the first row is the solution of the optimization problem.

## 5.3 Generalizations of the Notion of Convex Set

One of the reasons that the idea of convex set is so powerful is that it is rather rigid. For that reason, several generalizations of the concept have been developed over the years. We shall briefly treat some of them here.

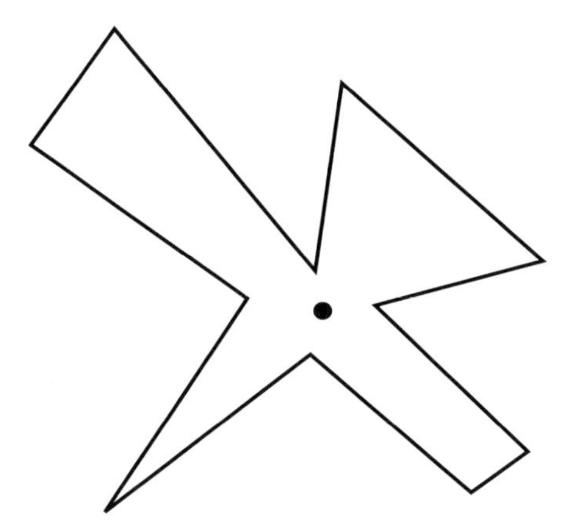

Figure 5.6: A star convex domain (that is not convex).

**Definition 5.15** We say that a domain $\Omega \subseteq \mathbb{R}^N$ is *starlike* or *star convex* if there is a distinguished point $X \in \Omega$ (called the *star point*) so that, if $P \in \Omega$, then segment $\overline{XP}$ lies in $\Omega$.

Plainly any convex domain is star convex. The converse is not true, however. See Figures 5.6 and 5.7 for some illustrations of the concept.

It is straightforward to verify that the Minkowski sum of two star convex domains is star convex. Also the set-theoretic product of two star convex domains is star convex. One may verify that a star convex domain is contractible, hence simply connected.

Unfortunately this is not true for either the union or the intersection—see the example below. As a result, we do not consider the star convex hull of a set. Since the process of taking the convex hull is one of the most important in convexity theory, this limits the utility of the new idea.

**Example 5.16** Consider the two domains shown in Figures 5.8 and 5.9. Each is star convex. Yet the union, shown in Figure 5.10, is plainly *not* star convex. And the intersection (Figure 5.11) is disconnected, so certainly not start convex. ∎

Another standard generalization of the standard notion of convexity is orthogonal convexity. A domain $\Omega \subseteq \mathbb{R}^N$ is *orthogonally convex* or *ortho-convex* if any segment parallel to one of the coordinate axes that connects two points of $\Omega$ lies entirely in $\Omega$.

The Minkowski sum of two ortho-convex sets is certainly ortho-convex. So is the product. It is straightforward to check that the intersection of ortho-convex sets is still ortho-convex. So we may consider the ortho-convex hull

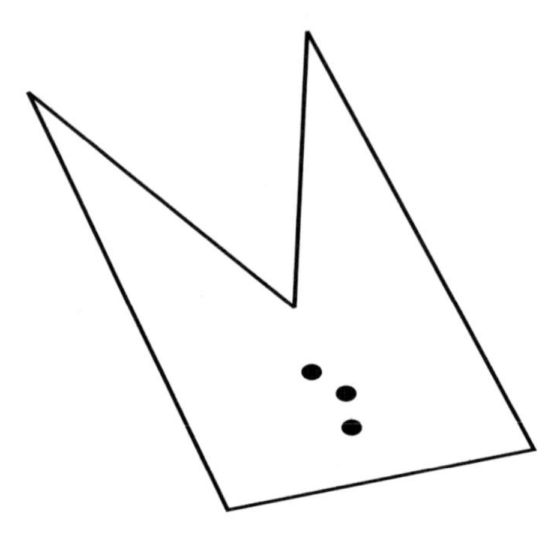

Figure 5.7: A star convex domain with several star points.

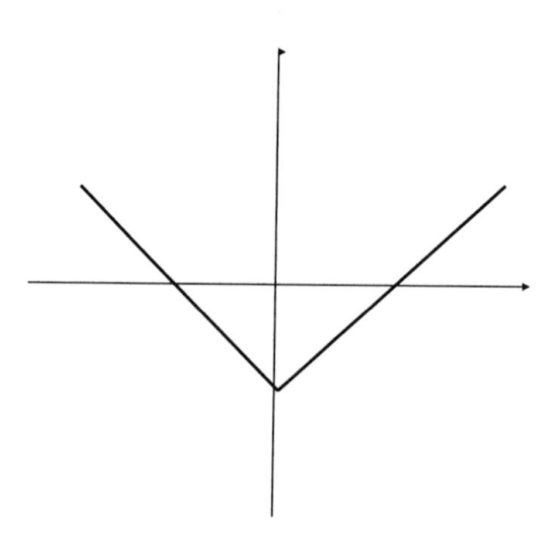

Figure 5.8: A star convex domain.

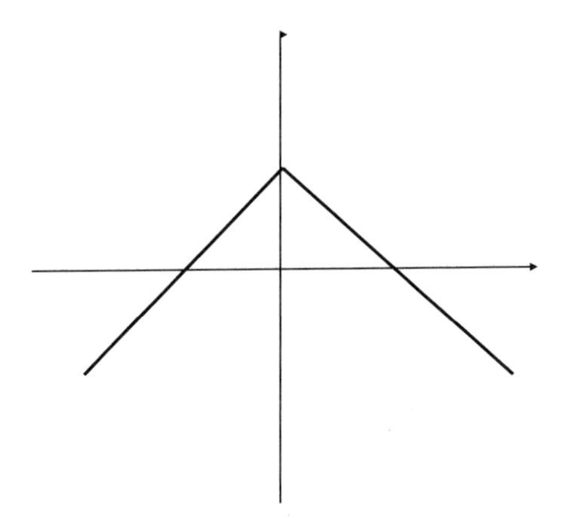

Figure 5.9: Another star convex domain.

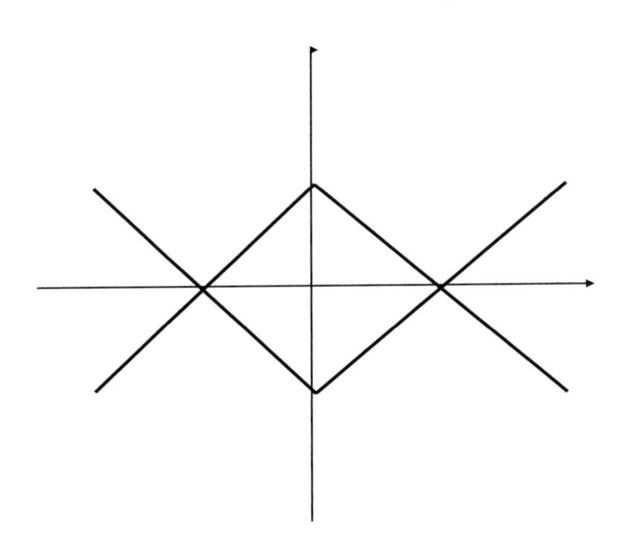

Figure 5.10: The union is not star convex.

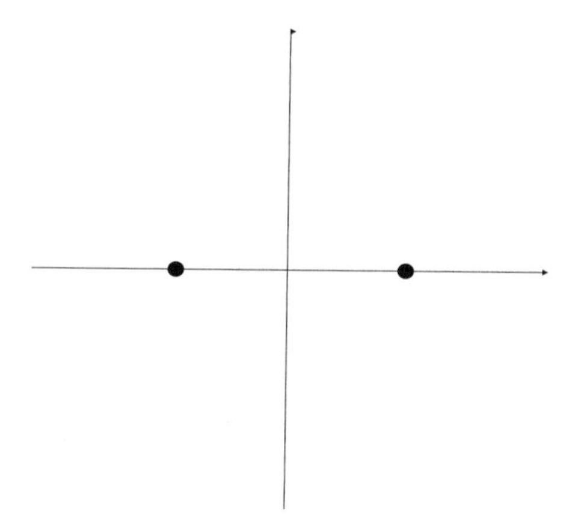

Figure 5.11: The intersection is not star convex.

of a given domain. However (see the example below) the union of two ortho-convex domains need not be ortho-convex.

**Example 5.17** Consider the domain shown in Figure 5.12 and the domain shown in Figure 5.13. The union of these two domains is exhibited in Figure 5.14, and it is certainly not ortho-convex. ∎

**Example 5.18** The set shown in Figure 5.15 has convex hull a solid triangle as shown in Figure 5.16. But its ortho-convex hull is just the set itself. ∎

It is easy to see that a star convex domain is rather natural. We may use the star point as the origin of coordinates. In at least some circumstances the star point will be the center of mass of the domain. Ortho-convex domains are natural when one is tied to a particular coordinate system.

There are many generalizations of the notion of convexity. These come up, in particular, in the context of optimization theory. Quasi-convex functions and pseudo-convex functions are also important in this context. A good reference is [ADSZ].

## 5.4 An Integral Representation for Convex Functions

In this section we derive an integral formula for a convex function that exhibits explicitly that the convex function is the supremum of its linear support functions. For that reason the formula is quite natural, and can be rather useful. For a reference, we offer [HOR], which is a useful book for many purposes.

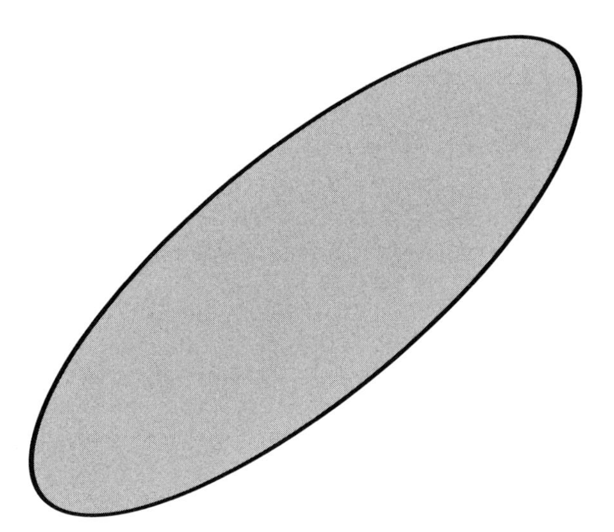

Figure 5.12: An ortho-convex domain.

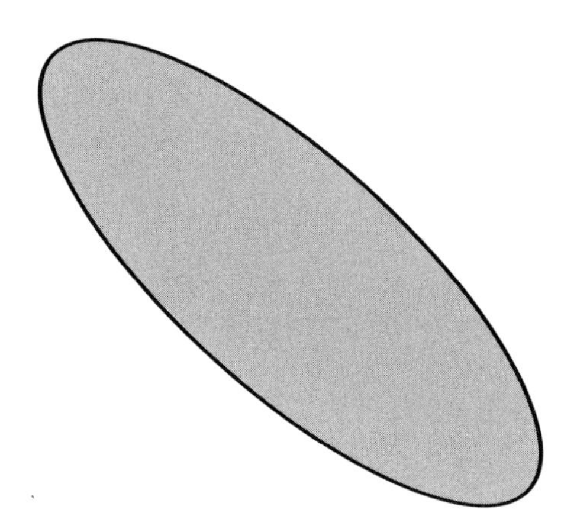

Figure 5.13: Another ortho-convex domain.

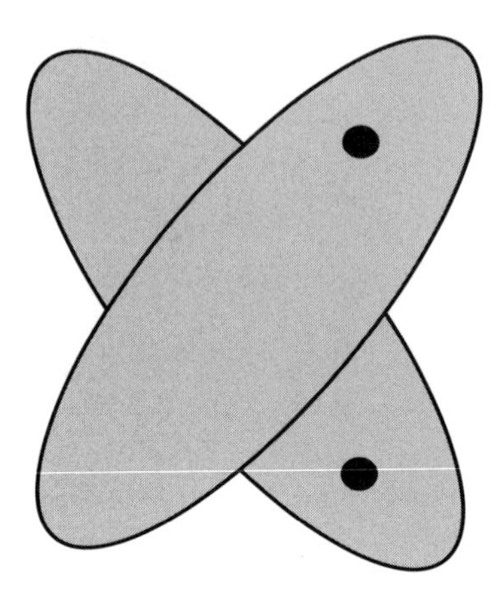

Figure 5.14: The union is not orthoconvex.

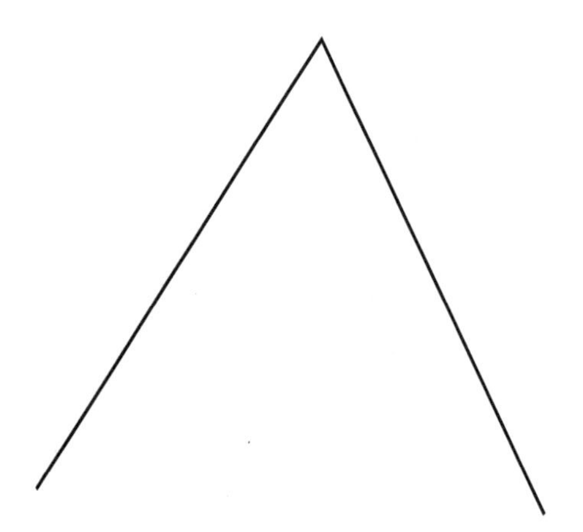

Figure 5.15: A nonconvex set.

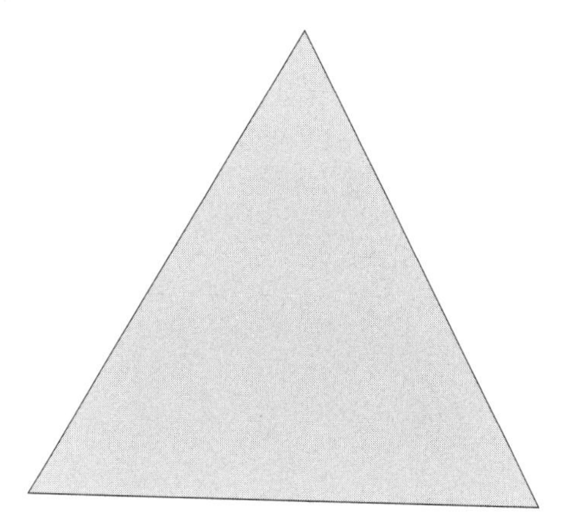

Figure 5.16: The convex hull.

Let $f$ be a bounded, convex function on the bounded, open interval $I = (a, b)$. Then of course $f$ can be extended continously to the endpoints (just by the monotonicity of the first derivative). We define

$$G_I(x,y) = \begin{cases} \dfrac{(x-b)(y-a)}{(b-a)} & \text{if} \quad a \le y \le x \le b \\ \dfrac{(x-a)(y-b)}{(b-a)} & \text{if} \quad a \le x \le y \le b \end{cases}$$

for $x, y \in \overline{I}$. We note that $G_I$ has the following properties:

- $G_I$ is continuous;
- $G_I \le 0$ on $\overline{I} \times \overline{I}$;
- $G_I = 0$ on $\partial(\overline{I} \times \overline{I})$;
- $G_I$ is symmetric in $x$ and $y$;
- $G_I$ is affine in each variable with the other variable fixed;
- On the diagonal, the derivative has a jump that equals 1.

We conclude, in particular, that $G_I$ is convex in each variable separately.

THEOREM 5.19 *There is a uniquely determined positive measure $\mu$ on $I$ such* *that*

$$f(x) = \int_I G_I(x,y)d\mu(y) + \frac{x-a}{b-a}f(b) + \frac{x-b}{a-b}f(a) \quad \text{for } x \in \overline{I}\,. \tag{5.19.1}$$

*Thus we have in particular that*

$$\int_I (x-a)(b-x)\, d\mu(x) < \infty\,. \tag{5.19.2}$$

*Conversely, for each positive measure $\mu$ satisfying (5.19.2), the integral in (5.19.1) defines a continuous, convex function $f$ in $\overline{I}$ which vanishes on $\partial I$, and so that $f'' = d\mu$ in the sense of distributions.*

**Proof:** It is convenient to first suppose that $f$ is defined and convex in a neighborhood of $\overline{I}$. Then

$$\mu(x) = f'_r(x)$$

(where the right-hand side denotes the right derivative of $f$ at $x$) is an increasing function in $\overline{I}$. Since, as we have already noted, $G_I = 0$ on the boundary, we may now integrate by parts to determine that

$$\begin{aligned}
\int_I G_I(x,y)\, d\mu(y) &= -\int_I \frac{\partial G_I(x,y)}{\partial y}\mu(y)\, dy \\
&= -\frac{x-b}{b-a}\int_a^x f'_r(y)\, dy \\
&\quad -\frac{x-a}{b-a}\int_x^b f'_r(y)\, dy \\
&= -\frac{x-b}{b-a}[f(x)-f(a)] \\
&\quad -\frac{x-a}{b-a}[f(b)-f(x)]\,.
\end{aligned}$$

That proves (5.19.1).

If we let $x = (a+b)/2$ and define $\delta(y) = (b-a)/2 - [y-(a+b)/2]$ (which is just the distance of $y$ to the boundary of $I$), then line (5.19.1) tells us that

$$\int_I \delta(y)\, d\mu(y) = f(a) + f(b) - 2f\left(\frac{1}{2}(a+b)\right)\,.$$

If we now drop the extra hypothesis that $f$ is convex in a neighborhood of $\overline{I}$, we can apply this last line to the interval $(a-\epsilon, b+\epsilon)$ for small $\epsilon > 0$. As

$\epsilon \to 0^+$ we may conclude that this displayed line holds for the interval $I$ itself, and that implies (5.19.2). The same argument shows that (5.19.1) is valid.

If $\varphi$ is a testing function in $I$, then we know that

$$\varphi(x) = \int_I G_I(x,y)\varphi''(y)\,dy;$$

this is true without any convexity assumption about $\varphi$. If $\mu$ is given and $f$ is defined by (5.19.1) with $\mu$ a positive measure satisfying (5.19.2), then the convexity of $G_1$ implies that $f$ is convex and

$$\begin{aligned} \int_I f(x)\varphi''(x)\,dx &= \iint_{I\times I} G_I(x,y)\varphi''(x)\,dx d\mu(y) \\ &= \int_I \varphi(y)\,d\mu(y)\,. \end{aligned}$$

The symmetry of $G_I(x,y)$ has obviously played a role here. Thus $d\mu = f''$ in the sense of distributions. That establishes the uniqueness. $\square$

## 5.5 The Gamma Function and Convexity

The Gamma function, to be defined momentarily, is one of the most important special functions of classical mathematical analysis. It is of course a generalization of the factorial function, it plays a role in the proof of Stirling's formula, and it has profound connections with the Riemann zeta function. In this section we explore some properties of the Gamma function that relate to convexity properties. This is a section about hard analysis, but it is fascinating material.

Recall that the Gamma function is defined by

$$\Gamma(z) = \int_0^\infty e^{-t}t^{z-1}\,dt\,.$$

We use $z$ to denote the variable because it is often convenient to think of a complex variable. Certainly the integral defining the Gamma function converges when $\operatorname{Re} z > 0$.

It is a standard fact, easily proved with integration by parts, that

$$\Gamma(z+1) = z\Gamma(z) \tag{5.20}$$

for $\mathrm{Re}\, z > 0$. Hence $\Gamma(n+1) = n!$, and the Gamma function extends the factorial function. Also

$$\Gamma(1) = 1\,. \tag{5.21}$$

The Gamma function is *not* characterized by these last two properties. See [HOR] for discussion of this point. But, by bringing convexity into the picture, we can come up with a characterization. It is as follows.

THEOREM 5.22 *The function* $\log \Gamma$ *is a convex function on the positive real axis, and there is no other positive function satisfying (5.20) and (5.21) that has this property.*

Given the important nature and canonical role that the $\Gamma$ function has, a uniqueness result of this kind is both of philosophical and practical importance.

The proof of this result is quite interesting, just because it entails several other notable and useful analytic characterizations of convex functions. We devote the rest of this section to proving the theorem, recording these ancillary results along the way:

LEMMA 5.23 *Let* $f$ *be a real-valued function defined on an interval* $I$. *Let* $\varphi$ *be another function defined on an interval* $J$ *and taking values in* $I$. *Then*

**(a)** *The function* $f \circ \varphi$ *is convex for every convex* $f$ *if and only if* $\varphi$ *is affine;*

**(b)** *The function* $f \circ \varphi$ *is convex for every convex* $\varphi$ *if and only if* $f$ *is convex and increasing.*

**Proof:** If $f \circ \varphi$ is convex for the choice $f(x) = x$ and also for the choice $f(x) = -x$, then $\varphi$ is clearly both convex and concave. Hence it is affine. Conversely, if $\varphi$ is affine then clearly $f \circ \varphi$ inherits convexity from $f$ (either by the chain rule, or just by thinking geometrically). That takes care of **(a)**.

Now assume that $f \circ \varphi$ is convex for every convex choice of $\varphi$. Taking $\varphi(x) = x$ we see immediately that $f$ is convex. Now if $x_1 < x_2$ are points of $I$, then set $\varphi(x) = x_1 + (x_2 - x_1)|x|$. Certainly this $\varphi$ is convex on the interval $[-1, 1]$. If we assume that $f \circ \varphi$ is convex then notice that $f \circ \varphi(\pm 1) = f(x_2)$ and $f \circ \varphi(0) = f(x_1)$. As a result, $f(x_1) \le f(x_2)$, so $f$ is increasing. Conversely, if $f$ is increasing and convex, then select $x_1, x_2 \in I$ and $\lambda_1$, $\lambda_2$ nonnegative with $\lambda_1 + \lambda_2 = 1$. Then we have

$$\begin{aligned} f(\varphi(\lambda_1 x_1 + \lambda_2 x_2)) &\le f(\lambda_1 \varphi(x_1) + \lambda_2 \varphi(x_2)) \\ &\le \lambda_1 f(\varphi(x_1)) + \lambda_2 f(\varphi(x_2))\,. \end{aligned}$$

Here the first inequality holds because $\varphi$ is convex and $f$ increasing; the second inequality holds since $f$ is convex. So $f \circ \varphi$ is convex and the proof is complete. □

PROPOSITION 5.24 *Let $g$ be a positive function defined on an interval $I$. Then $\log g$ is convex (resp. affine) on $I$ if and only if the function $t \mapsto e^{ct} g(t)$ is convex (resp. constant) on $I$ for every (resp. some) $c \in \mathbb{R}$.*

**Proof:** Assume that $\log g$ is convex. Then $t \mapsto ct + \log g(t)$ is convex. Since $u \mapsto e^u$ is convex and increasing, we may then apply the lemma to see that $t \mapsto e^{ct} g(t)$ is convex.

Conversely, assume that $t \mapsto e^{ct} g(t)$ is convex for every choice of $c$. Let $J \subset I$ be a compact subinterval. Then the maximum of $e^{ct} g(t)$ for $t \in J$ is assumed when $t \in \partial J$. Hence the maximum of $ct + \log g(t)$ is assumed when $t \in \partial J$. Thus $\log g$ is convex.

The condition for $\log g$ to be affine is trivial.

The proof is complete. □

**Proof of Theorem 5.22:** We may use the lemma to verify that $\log \Gamma$ is convex:

$$\frac{d^2}{dx^2} \left( e^{cx} \Gamma(x) \right) = \int_0^\infty e^{-t} (c + \log t)^2 e^{cx} t^{x-1} \, dt > 0 \, .$$

For uniqueness, we note that equation (5.20) (for the real variable $x$) can be written as

$$\log \Gamma(x+1) - \log \Gamma(x) = \log x \, .$$

Notice that the right-hand side is a concave function of $x$ that is $o(x)$ as $x \to +\infty$.

Our result then follows from the following proposition:

PROPOSITION 5.25 *Let $h(x)$ be a concave function on $\{x \in \mathbb{R} : x > 0\}$ such that $h(x)/x \to 0$ as $x \to +\infty$. Then the difference equation*

$$g(x+1) - g(x) = h(x) \tag{5.25.1}$$

*for $x > 0$ has one and only one convex solution $g$ with $g(1) = 0$. It is in fact given by*

$$g(x) = -h(x) + \lim_{n \to \infty} \left( x h(n) + \sum_{j=1}^{n-1} \left( h(j) - h(x+j) \right) \right) . \tag{5.25.2}$$

**Proof of the Proposition:** Notice that, for $x$ fixed, $[h(x+y)-h(x)]/y$ is decreasing and tends to 0 as $y \to +\infty$. If $y > 0$, then

$$\frac{h(x+y)-h(x)}{y} \leq \frac{h(x)-h(x/2)}{x/2} \to 0 \quad \text{as } x \to +\infty\,. \tag{5.25.3}$$

So we see that the sequence $a_n = h(n+1)-h(n)$ is decreasing and converges to 0 as $n \to \infty$.

Now equation (5.25.1) and the condition $g(1) = 0$ tell us that

$$\begin{aligned} g(x) + h(x) - xh(n) - \sum_{j=1}^{n-1} [h(j) - h(x+j)] & \\ = g(x+n) - g(n) - xh(n)\,. & \end{aligned} \tag{5.25.4}$$

If $g$ is convex and $k$ is an integer with $x \leq k$, then

$$\begin{aligned} h(n-1) &= g(n) - g(n-1) \\ &\leq \frac{g(x+n)-g(n)}{x} \\ &\leq \frac{g(k+n)-g(n)}{k} \\ &= \frac{h(n)+\cdots+h(n+k-1)}{k}\,. \end{aligned}$$

Therefore

$$\begin{aligned} -xa_{n-1} &\leq g(x+n) - g(n) - xh(n) \\ &\leq x \cdot \frac{(k-1)a_n + (k-2)a_{n+1} + \cdots + a_{n+k-2}}{k}\,, \end{aligned}$$

and this proves that $g(x+n) - g(n) - xh(n) \to 0$ as $n \to +\infty$. It follows then from (5.25.4) that $g$ must be of the form (5.25.2). This uniqueness by itself suffices to prove our theorem. But, in order to complete the proof of the proposition, we must also show that (5.25.2) converges to a convex function with $g(1) = 0$ and satisfying (5.25.1).

It is in fact clear that $g(1) = 0$ and that (5.25.1) will follow since $h(n) - h(x+n) \to 0$ as $n \to \infty$ by (5.25.3). Only the convergence requires some thought.

We write the limit (5.25.2) as $\sum_0^\infty u_n(x)$, where $u_0(x) = xh(1) - h(x)$ and

$$u_n(x) = x\,[h(n+1) - h(n)] + h(n) - h(x+n)\,, \quad n > 0\,.$$

Clearly each $u_n$ is convex. Because of the concavity of $h$, we see when $k \geq x$ that

$$h(n-1) - h(n) \leq \frac{h(n) - h(x+n)}{x} \leq \frac{(h(n) - h(k+n)}{k},$$

hence

$$\begin{aligned} x(a_n - a_{n-1}) &\leq u_n(x) \\ &\leq x\left(a_n - \frac{a_n + \cdots + a_{n+k-1}}{k}\right) \end{aligned}$$

for $n > 0$.

Thus $u_n \geq 0$. Since

$$\sum_{n=1}^{\infty}\left(a_n - \frac{a_n + \cdot a_{n+k-2}}{k}\right) = \sum_{j=1}^{k} a_j\left(1 - \frac{j}{k}\right) < \infty,$$

the convergence of $\sum_0^\infty u_n(x)$ follows. Also the sum is convex since each summand is. The proof is complete. □

We conclude this section by noting that the formula (5.25.2) applied to the function $g(x) = \log\Gamma(x)$ yields the very useful product formula

$$\Gamma(x) = \lim_{n\to\infty} \frac{n^x n!}{x(x+1)\cdot\cdots\cdot(x+n)}.$$

So far in this chapter we have been able to relate analytic convexity to a number of classical topics in analysis—ranging from special functions to optimization to integral formulas. This has been one of the main purposes of the book. In the next section we re-discover Jensen's inequality, Höder's inequality, Minkowski's inequality, and many other classical facts.

## 5.6 Hard Analytic Facts

In this section we collect a number of basic hard analytic facts about convex functions. The reference [HOR] is a good source for further information about these matters.

PROPOSITION 5.26 *Let $f_1, f_2, \ldots, f_k$ be convex functions on an interval $I$. Let $c_1, c_2, \ldots, c_k$ be nonnegative real constants. Then $f = \sum_j c_j f_j$ is also convex.*

**Proof:** We calculate directly that

$$\begin{aligned} f((1-t)x+ty) &= \sum_{j=1}^{k} c_j f_j((1-t)x+ty) \\ &\leq \sum_{j=1}^{k} (1-t)f_j(x) + tf_j(y) \\ &= (1-t)\sum_{j=1}^{k} f_j(x) + t\sum_{j=1}^{k} f_j(y) \\ &= (1-t)f(x) + tf(y)\,. \end{aligned}$$

That completes the proof. □

PROPOSITION 5.27 *Let $\{f_\alpha\}_{\alpha\in A}$ be convex functions on an interval $I$. Define*

$$J = \{x \in I : f(x) \equiv \sup_{\alpha\in A} f_\alpha(x) < \infty\}\,.$$

*Then $J$ is an interval and $f$ is a convex function on $J$. Note that, in this case, $J$ may be empty.*

*Second, let $f_j$ be convex functions on an interval $I$. Define*

$$J = \{x \in I : F(x) \equiv \limsup_{j\to\infty} f_j(x) < \infty\}\,.$$

*Then $J$ is an interval and $F$ is a convex function on $J$ unless $F \equiv -\infty$ in the interior of $J$ or $J$ is a singleton.*

**Proof:** For the first assertion, notice that if $x < y$ are elements of $J$ and $0 \leq t \leq 1$ then, for each $j$,

$$f_j((1-t)x+ty) \leq (1-t)f_j(x) + tf_j(y)$$

so that

$$f_j((1-t)x+ty) \leq (1-t)f(x) + tf(y)\,.$$

Taking the supremum over $j$ now yields that

$$f((1-t)x+ty) \leq (1-t)f(x) + tf(y) < \infty\,.$$

That shows that $J$ is an interval and that $f$ is convex on $J$.

For the second assertion, assume that $J$ contains two distinct points $x$ and $y$ and that $F$ takes a finite value at each of those points. Let $0 \leq t \leq 1$. Let $\epsilon > 0$ and choose $j$ so large that $f_j(x) < F(x) + \epsilon$ and $f_j(y) < F(y) + \epsilon$. Then

$$\begin{aligned} f_j((1-t)x + ty) &\leq (1-t)f_j(x) + tf_j(y) \\ &\leq (1-t)[F(x) + \epsilon] + t[F(y) + \epsilon] \,. \end{aligned}$$

As $j \to \infty$ along any subsequence, we find that the subsequential limit is dominated by $(1-t)[F(x) + \epsilon] + t[F(y) + \epsilon]$. Hence

$$F((1-t)x + ty) \leq (1-t)[F(x) + \epsilon] + t[F(y) + \epsilon] \,.$$

Therefore $F$ is finite at $(1-t)x + ty$ and $F$ is convex.

If $J$ only contains one point then we cannot argue as in the previous paragraph. If $F$ takes the value $-\infty$ at any point of $J$ then it is also the case that the preceding paragraph is invalid.

That completes the proof. $\square$

We take a moment now to reformulate the definition of convexity for a function. The original definition

$$f((1-t)x + ty) \leq (1-t)f(x) + tf(y)$$

can be restated as

> The function $f$ on the interval $I$ is convex if the graph lies below the chord between any two points of the graph. That is, for every compact interval $J \subset I$ and every linear function $\mathcal{L}$, we have
>
> $$\sup_J (f - \mathcal{L}) = \sup_{\partial J} (f - \mathcal{L}) \,.$$
>
> See Figure 5.17.

THEOREM 5.28 *The real-valued function $f$ on the interval $I$ is convex if and only if, for every $x \in I$, the difference quotient*

$$\frac{f(x+h) - f(x)}{h}$$

*is an increasing function of $h$ whenever $x + h \in I$ and $h \neq 0$.*

**Proof:** Our usual definition of convex function can be rewritten as follows. Let $x_1 < x < x_2$ be points of $\mathbb{R}$. Then

$$x = \lambda_1 x_1 + \lambda_2 x_2$$

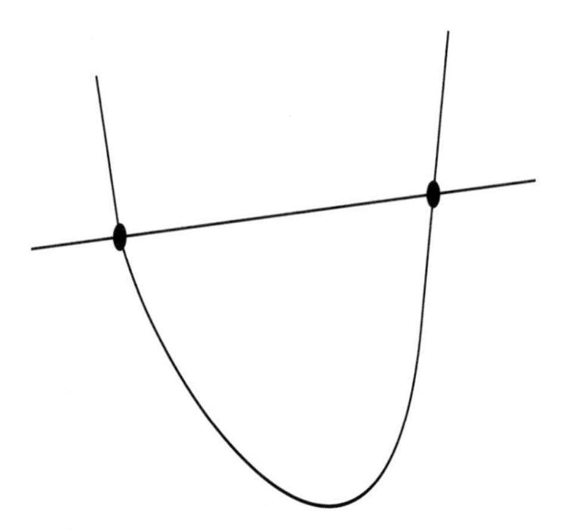

Figure 5.17: Another look at convexity.

with

$$\lambda_1 = \frac{x_2 - x}{x_2 - x_1}$$

and

$$\lambda_2 = \frac{x - x_1}{x_2 - x_1}.$$

Thus $f(x) = f(\lambda_1 x_1 + \lambda_2 x_2) \leq \lambda_1 f(x_1) + \lambda_2 f(x_2)$ can be rewritten as

$$(x_2 - x_1)f(x) \leq (x_2 - x)f(x_1) + (x - x_1)f(x_2)$$

or

$$\frac{f(x) - f(x_1)}{x - x_1} \leq \frac{f(x_2) - f(x)}{x_2 - x}. \tag{5.28.1}$$

That is the desired result. □

PROPOSITION 5.29 *If $f$ is convex on an interval $I$, then the left derivative $f'_\ell(x)$ and the right derivative $f'_r(x)$ exist at every interior point $x$ of $I$. They are both increasing functions.*

*If $x_1 < x_2$ are in the interior of $I$, then we have*

$$f'_\ell(x_1) \leq f'_r(x_1) \leq \frac{f(x_2) - f(x_2)}{x_2 - x_1} \leq f'_\ell(x_2) \leq f'_r(x_2). \tag{5.29.1}$$

*In particular, $f$ is Lipschitz continuous in every compact subinterval $J$ contained in the interior of $I$.*

**Proof:** This is immediate from the preceding theorem. □

Now (5.29.1) of the last proposition implies, for $x_1 < x_2$ in $I$, that

$$\lim_{\epsilon\to 0+} f_r'(x_1+\epsilon) \le \frac{f(x_2)-f(x_1)}{x_2-x_1} \le \lim_{\epsilon\to 0+} f_\ell'(x_2-\epsilon)\,. \tag{5.29.2}$$

If we now let $x_2$ decrease to $x_1$ or $x_1$ increase to $x_2$ we then obtain

THEOREM 5.30 *If $f$ is convex in the interval $I$ and $x$ is an interior point of $I$, then*

$$f_r'(x) = \lim_{\epsilon\to 0+} f_r'(x+\epsilon) = \lim_{\epsilon\to 0+} f_\ell'(x+\epsilon)$$

*and*

$$f_\ell'(x) = \lim_{\epsilon\to 0+} f_r'(x-\epsilon) = \lim_{\epsilon\to 0+} f_\ell'(x-\epsilon)\,.$$

*We shall therefore write $f'(x+0) = f_r'(x)$ and $f'(x-0) = f_\ell'(x)$.*

*The following conditions are equivalent:*

**(1)** *$f_\ell'$ is continuous at $x$;*

**(2)** *$f_r'$ is continuous at $x$;*

**(3)** *$f_r'(x) = f_\ell'(x)$, that is to say, $f$ is differentiable at $x$.*

*These equivalent conditions are fulfilled at all but countably many points.*

**Proof:** This result is immediate from Proposition 5.29. The very last statement follows from the fact that, if $x_1 < x_2$ are elements of $I$, then

$$\sum_{x_1<x<x_2} \left[f_r'(x) - f_\ell'(x)\right] \le f_\ell'(x_2) - f_r'(x_1) < \infty\,.$$ □

PROPOSITION 5.31 *Let $x_1 < x_2 < \cdots$ be real numbers and let $a_j > 0$ be selected so that $\sum_j a_j(1+|x_j|) < \infty$. Then*

1. *$f(x) \equiv \sum_j a_j|x-x_j|$ is a convex function;*
2. *$f_r'(x_j) - f_\ell'(x_j) = 2a_j$ for all $j$;*
3. *$f'(x) = \sum_j a_j \operatorname{sgn}(x-x_j)$ for all $x \ne x_j$, all $j$.*

**Proof:** Clearly the series defining $f$ is convergent. Also $f$ is the sum of convex functions, so it is convex.

Now fix an index $j$. For $\ell \neq j$, the left and right derivatives of $|x - x_\ell|$ at $x_j$ are equal. So, in studying these one-sided derivatives, we may concentrate on $|x - x_j|$. But it is obvious that the left derivative is $-1$ and the right derivative is 1. That gives the second statement.

The third statement is proved with a similar analysis. $\square$

THEOREM 5.32 *If $f_j$ are convex on an open interval $I$ and $f_j \to f$ pointwise then $f_j \to f$ uniformly on compact sets. Assume that $f$ is continuous. Then the limit function $f$ is convex.*

**Proof:** Take the compact set to be $[0, 1]$ lying in the open interval $I$. We proceed in several steps.

**The limit function is convex:** This assertion is proved simply by passing to the limit in the definition of convex function.

**The $f_j$ are uniformly bounded** : We know that $\{f_j(0)\}$ and $\{f_j(1)\}$ converge. Hence it is clear that $\{f_j(0)\}$ and $\{f_j(1)\}$ are bounded above. So, by convexity, the functions $\{f_j\}$ are bounded above. If the $\{f_j\}$ were not bounded below then the sequence $\{f_j\}$ could not converge pointwise to a limit function $f$. We conclude that $\{f_j\}$ is bounded.

**There is a constant $K$ so that $|f_j(x) - f_j(y)| \leq K|x - y|$ for all $j$ and all $x, y \in I$:** Suppose not. Then there are $x_\ell, y_\ell \in I$ such that $|f_{j_\ell}(x_\ell) - f_{j_\ell}(y_\ell)| \geq \ell|x_\ell - y_\ell|$. Then there are subsequences $x_{\ell_k}$, $y_{\ell_k}$ that converge to limit points $x_0$, $y_0$ in $I$. Of necessity, $x_0$ and $y_0$ would have to be equal, and would have to equal one of the endpoints of $I$. But this would contradict the continuity of the limit function $f$.

**The functions $f_j$ converge uniformly:** Because the family $\{f_j\}$ is equibounded and equicontinuous, this is just an application of the Ascoli-Arzela theorem (see [RUD]). But in fact we can apply the Ascoli-Arzela theorem to see that any subsequence of $\{f_j\}$ has a subsequence that converges uniformly. That implies that the full sequence $\{f_j\}$ converges uniformly. $\square$

PROPOSITION 5.33 *Let $f$ be convex on the interval $I$, and assume that $f$ is bounded above. If $I$ is unbounded on the right, then $f$ is nonincreasing. If $I$ is unbounded on the left, then $f$ is nondecreasing.*

**Proof:** If $f$ is increasing at any point $x_0$, then the derivative is positive—say that $f'(x_0) = c > 0$. But the first derivative of a convex function is increasing from left to right, so the function would satisfy $f'(x) \geq c$ for $x \geq c$. Thus, by the fundamental theorem of calculus, $f$ would be unbounded if $I$ is unbounded on the right. As a result, the $f$ satisfying the hypothesis of the proposition is nonincreasing.

A similar argument applies when $I$ is unbounded on the left. □

Next we look at a form of the mean value theorem that is useful for our studies.

PROPOSITION 5.34 *Let $f$ be a continuous function on the interval $I = \{x \in \mathbb{R} : a \leq x \leq b\}$. Assume that $f'_r(x)$ exists when $a \leq x < b$. If $f'_r(x) \geq C$ for all such $x$, then $f(b) - f(a) \geq C(b-a)$. If instead $f'_r(x) \leq C$, then $f(b) - f(a) \leq C(b-a)$.*

**Proof:** It is enough to prove the first of these statements. If $C' < C$, then define

$$F = \{x \in I : f(x) - f(a) \geq C'(x-a)\} .$$

Notice that $F$ is closed because $f$ is continuous. Also $a \in F$. So $F$ contains its supremum $\beta$, and in fact that supremum equals $b$. For, if not then, for small $h > 0$,

$$\begin{aligned} f(\beta + h) - f(a) &= [f(\beta + h) - f(\beta)] + [f(\beta) - f(a)] \\ &\geq C'h + C'(\beta - a) = C'(\beta + h - a) . \end{aligned}$$

This contradicts the definition of $\beta$. As a result, $f(b) - f(a) \geq C'(b-a)$ for every $C' < C$, and that proves the result. □

This last result gives the inequality

$$\inf_{[a,b)} f'_r \leq \frac{f(b) - f(a)}{b-a} \leq \sup_{[a,b)} f'_r . \tag{5.35}$$

Now we have:

THEOREM 5.36 *If $f$ is a continuous function on the interval $I$, so that $f'_r$ exists at every interior point of $I$ and $f'_r$ increases with $x$, then $f$ is convex. Also*

$$\int_x^y f'_r(t)\,dt = f(y) - f(x)$$

*for all $x, y \in I$.*

*A similar result is true with $f'_\ell$ replacing $f'_r$.*

**Proof:** Of course we shall use the mean value theorem (and its consequences) above.

First notice that (5.35) immediately implies (5.29.2) when $x_1, x_2$ are interior points of $I$. By continuity, the inequality (5.29.2) persists at the endpoints. The convexity now follows in view of Theorem 5.32. The second statement follows because the right derivative of $\int_x^y f'_r(t)\,dt$ with respect to $y$ is equal to $f'_r(y)$ by monotonicity and right continuity. □

An immediate remark is that the characterization of convexity of a function by the nonnegativity of the second derivative follows from the last result.

**Example 5.37** The function $f_a(x) = e^{ax}$ is convex on $\mathbb{R}$ for any possible fixed choice of $a \in \mathbb{R}$. If $r \geq 1$, then $g_r(x) \equiv x^r$ is convex for nonnegative $x$. If $r < 0$, then $g_r$ is convex for $x \geq 0$. If $0 < r \leq 1$, then $g_r$ is concave for $x \geq 0$.

The functions $h(x) = x \log x$ and $k(x) = -\log x$ are convex for positive $x$.

■

Another elementary, but useful and conceptually satisfying, property of convex functions is this:

PROPOSITION 5.38 *Convexity is a local property. This means that, if $f$ is defined in an interval $I$ and each $x \in I$ has a neighborhood $J$ so that $f$ is convex on $J$, then $f$ is convex on $I$.*

Our most basic definition of convex function is in terms of affine majorants. But there is also a definition in terms of affine minorants:

PROPOSITION 5.39 *A real-valued function $f$ defined on an interval $I$ is convex if and only if, for each $x$ in the interior of $I$, there is an affine function $\tau$ with $\tau \leq f$ and $\tau(a) = g(x)$. See Figure 5.18.*

**Proof of the Proposition:** First assume that $f$ is convex. Choose $s \in [f'_\ell(x), f'_r(x)]$. Define $g(y) = f(x) + s(y - x)$. Since $g(x) = f(x)$ and formula (5.35) gives both

$$f(y) \geq f(x) + (y - x)f'_r(x) \geq g(x) \quad \text{for } y \geq x;$$

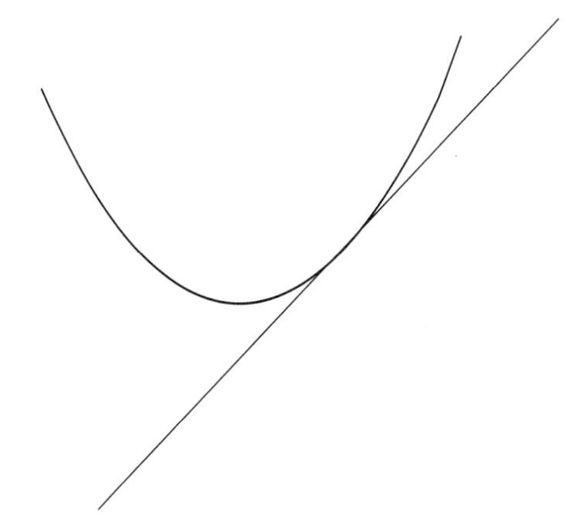

Figure 5.18: Characterization of a convex function in terms of minorants.

$$f(y) \geq f(x) + (y - x)f'_\ell(x) \geq g(y) \quad \text{for } y \leq x \,,$$

we see that the necessity is established.

Now assume that $f$ satisfies the affine support function condition enunciated in the theorem. We must then prove the usual definition of convexity. We may take $x_1, x_2$ as distinct elements of $I$ and define $x = \lambda_1 x_1 + \lambda_2 x_2$ an interior point of $I$. If now $\tau$ is an affine minorant of $f$ with $\tau(x) = f(x)$ then we have

$$\begin{aligned} \sum_1^2 \tau_j f(x_j) &\geq \sum_1^2 \lambda_j \tau(x_j) \\ &= g\left(\sum_1^2 \lambda_j x_j\right) \\ &= g(x) - f(x) \,. \end{aligned}$$

That establishes the result. □

Theorem 5.40 (Jensen) *Let $f$ be a convex function on the interval $J$. Let $\alpha$ be an integrable function on the unit interval $I = [0,1]$ with values in $J$. Then*

$$f\left(\int_0^1 \alpha(t)\, dt\right) \leq \int_0^1 f(\alpha(t))\, dt \,.$$

**Proof:** The result is obvious—with equality instead of inequality—when $f$ is affine. Now think of a general $f$ as the supremum of affine functions (as in the preceding result). □

We note in passing that Jensen's inequality is still true if Lebesgue measure on the unit interval is replaced by any positive measure on a compact space having total mass 1.

Of course the theory of convex functions is of considerable aesthetic and intrinsic interest. But it also has many useful applications. We begin with a simple one that is commonly referred to as the comparison of the arithmetic and geometric means.

PROPOSITION 5.41 *Let $a_j > 0$, $\gamma_j > 0$, $j = 1, \ldots, k$, and assume that $\sum \gamma_j = 1$. Then*

$$\prod_{j=1}^{k} a_j^{\gamma_j} \leq \sum_{j=1}^{k} \gamma_j a_j \,.$$

*The inequality is strict unless all the $a_j$ are equal.*

**Proof:** If we write $a_j = e^{\alpha_j}$ for each $j$, then our inequality becomes

$$\exp\left(\sum_{j=1}^{k} \gamma_j \alpha_j\right) \leq \sum_{j=1}^{k} \gamma_j \exp \alpha_j \,.$$

This inequality is an immediate application of Jensen's inequality (see the comment after the enunciation and proof of the theorem), with $f(x) = e^x$. □

THEOREM 5.42 *Let $a_j > 0$, $\gamma_j > 0$, $j = 1, \ldots, k$, and suppose that $\sum_j \gamma_j = 1$. Then*

$$\sum_j \gamma_j a_j \leq \left(\sum_j \gamma_j a_j^p\right)^{1/p}$$

*provided that $p > 1$. We have equality only if all the $a_j$ are equal.*

**Proof:** Raise both sides to the power $p$. Then note that the function $x \mapsto x^p$ is convex for $x > 0$. So Jensen applies. □

THEOREM 5.43 (HÖLDER) *Let $p > 1$, $p' > 1$ with $1/p + 1/p' = 1$. Assume that $a_j$, $b_j$ are positive real numbers, $j = 1, \ldots, k$. Then*

$$\sum_j a_j b_j \leq \left(\sum_j a_j^p\right)^{1/p} \cdot \left(\sum_j b_j^{p'}\right)^{1/p'} .$$

*We have equality if and only if the quantities $a_j^p / b_j^{p'}$ are all equal.*

**Proof:** We replace $\gamma_j$ in Theorem 5.42 with $\gamma_j / \sum_\ell \gamma_\ell$ for arbitrary $\gamma_j$. The result is

$$\sum_{j=1}^{k} \frac{\gamma_j}{\sum_\ell \gamma_\ell} a_j \leq \left(\sum_j \frac{\gamma_j}{\sum_\ell \gamma_\ell} a_j^p\right)^{1/p}$$

or

$$\sum_j \gamma_j a_j \leq \left(\sum_\ell \gamma_\ell\right)^{1/p'} \left(\sum_j \gamma_j a_j^p\right)^{1/p} .$$

Now let $\gamma_j = b_j^{p'}$ amd $a_j = a_j b_j^{1-p'}$. Then we have

$$\begin{aligned} \sum_j b_j^{p'} a_j b_j^{1-p'} &\leq \left(\sum_\ell b_\ell^{p'}\right)^{1/p'} \cdot \left(\sum_j b_j^{p'} a_j^p b_j^{p-p'p}\right)^{1/p} \\ &= \left(\sum_\ell b_\ell^{p'}\right)^{1/p'} \cdot \left(\sum_j a_j^p\right)^{1/p} . \end{aligned}$$

In conclusion,

$$\sum_j a_j b_j \leq \left(\sum_j b_j^{p'}\right)^{1/p'} \cdot \left(\sum_j a_j^p\right)^{1/p} .$$

That concludes the proof. □

Closely related to Hölder's inequality is Minkowski's inequality. We prove that result next.

THEOREM 5.44 (MINKOWSKI) *Let $a_1, a_2, \ldots, a_k$ and $b_1, b_2, \ldots, b_k$ be non-negative numbers and $p > 1$. Then*

$$\left(\sum_j (a_j + b_j)^p\right)^{1/p} \leq \left(\sum_j a_j^p\right)^{1/p} + \left(\sum_j b_j^p\right)^{1/p} .$$

*The inequality is strict unless $a = (a_1, a_2, \ldots, a_k)$ and $(b_1, b_2, \ldots, b_k)$ are linearly dependent.*

**Proof:** If $a = 0$ or $b = 0$ then the result is trivial. If neither is zero then we may choose constants $\alpha > 0$ and $\beta > 0$ so that

$$\alpha \left( \sum_j a_j^p \right)^{1/p} = \beta \left( \sum_j b_j^p \right)^{1/p} - 1 .$$

Then define a convex function of the variable $s$ by

$$f(s) = \sum_j (\alpha s a_j + \beta(1-s) b_j)^p .$$

We see that $f(0) = f(1) = 1$. The function is not affine on $[0,1]$ unless $\alpha a = \beta b$. Otherwise $f(s) < 1$ when $\alpha s = \beta(1-s)$. Thus $s = \beta/(\alpha+\beta)$ and $\alpha s = \alpha\beta/(\alpha+\beta)$. Therefore

$$\left( \sum_j (a_j + b_j)^p \right)^{1/p} < \frac{1}{\alpha} + \frac{1}{\beta} .$$

That completes the proof. □

## 5.7 Sums and Projections of Convex Sets

There are a number of contexts in mathematical physics in which sums and projections of convex sets arise. In particular, one wants to know whether the sum of two smoothly bounded convex sets is smoothly bounded, or whether the projection of a smoothly bounded convex set is smoothly bounded.

The answers to these questions are not at all obvious. In general they are "no," but there are plausible conditions that make the answers "yes." The answers depend on the dimension of the ambient space, the degree of convexity of the component domains, and other subtle considerations. We cannot explore all the byways here, but we shall begin by presenting some examples that put the matter in perspective. Some of the key papers in the subject are [KIS1], [KIS2], [KRP4].

It is an interesting fact that the problem of studying sums of sets and the problem of studing projections of sets are closely related. To see this, suppose that

$$F = \{(x,y) \in \mathbb{R}^2 : y > f(x)\}$$

and

$$G = \{(x, y) \in \mathbb{R}^2 : y > g(x)\}.$$

Then $F + G$ is the projection of the set

$$\{y > f(s) + g(t) : s + t = x\}.$$

Thus, in addition (see below), one sees that the boundary of $F + G$ is given by the infimal convolution $y > f \sqcap g$.

**Example 5.45** There is a domain in $\mathbb{R}^3$ with $C^\infty$ boundary such that the projection of the domain into $\mathbb{R}^2$ is only $C^2$.

A curious result of Kiselman is that, if the domain in $\mathbb{R}^3$ has real analytic boundary, then the projection will always have smoothness $C^{2+\epsilon}$. And the size of $\epsilon$ may be estimated.

In any event, let us now construct the above-mentioned example.

Let $u$ be a function on $\{y \in \mathbb{R} : y > 0\}$ defined by

$$u''(y) = \sin^2(1/y)\exp(-1/y).$$

Then define $u$ by double integration, specifying that $u'(0) = 0$ and $u'(y) > 0$ when $0 < y < 1/2$. We may take $u$ to be defined to be even on all of $\mathbb{R}$.

Now set

$$f(x, y) = x^2(4 - y + y^2/2) + u(y).$$

We may think of $f$ as a convex function on all of $\mathbb{R}^2$. Now set

$$\Omega = \{(x, y) : f(x, y) \leq z\}.$$

This is certainly a convex domain in $\mathbb{R}^3$. Let $\pi$ be the projection $\pi(x, y, z) = (x, z)$. Then $\pi(\Omega)$ has only $C^2$ boundary.

We note that $\pi(\Omega)$ is unbounded, but an appropriate compact subset of this domain will satisfy all our requirements. ∎

For the next example we need a new tool called the infimal convolution. If $f$ and $g$ are two real-valued functions on the real line then we define their *infimal convolution* by

$$f \sqcap g(x) = \inf\{f(y) + g(x - y) : x, y \in \mathbb{R}\}.$$

LEMMA 5.46 *Let $F$ be the epigraph of $f$ and $G$ be the epigraph of $g$. Then $F + G$ is the epigraph of $f \sqcap g$.*

**Proof:** Let $x$ be in the epigraph of $f \sqcap g$. Writing $x = (x_1, x_2)$, we then know that

$$x_2 \geq f \sqcap g(x_1)$$

or

$$x_2 \geq \inf\{f(y) + g(x_1 - y) : x, y \in \mathbb{R}\}.$$

But this says that, for $\epsilon > 0$ small,

$$x_2 \geq f(y) + g(x_1 - y) - \epsilon$$

for some $y \in \mathbb{R}$. Thus $p = (y, f(y)) \in F$ and $q = (x_1 - y, g(x_1 - y)) \in G$ so that $(x_1, f(y) + g(x_1 - y)) \in F + G$ and $x_2 \geq p_2 + q_2$. So $x \in F + G$.

The argument in reverse is similar, and we omit the details. □

**Example 5.47** Let $F = \{(x, y) : y \geq x^4/4 \equiv f(x)\}$ and $G = \{(x, y) : y \geq x^6/6 \equiv g(x)\}$. Then obviously $F$ and $G$ each has $C^\infty$ boundary (each is unbounded) but $F + G$ has only $C^{20/3}$ boundary.

To prove this contention, we of course use the infimal convolution.

In fact

$$f \sqcap g(x) = \inf\left(\frac{y^4}{4} + \frac{(x-y)^6}{6}\right) = \frac{x^6}{6} - \frac{3|x|^{20/3}}{4} + r(x),$$

where $r \in C^7(\mathbb{R})$. This last is seen to be true by noting that the infimum is attained at a unique point $y$ which satisfies $f'(y) = g'(x - y)$, that is $y^3 = (x - y)^5$. Thus $y = x^{5/3}$ is a reasonable approximation. If we accept this calculation, then we see that $f \sqcap g$ is $C^{20/3}$ but in no lesser Lipschitz class.

In summary, the convex sets $F$ and $G$ have real analytic boundaries, but their Minkowski sum has boundary of class $C^{20/3}$ only. ∎

# Exercises

**1.** Calculate the polar of the set

$$\Omega = \{(x_1, x_2) \in \mathbb{R}^2 : x_2 = x_1, 0 \leq x_1 \leq 1\}$$
$$\cup\{(x_1, x_2) \in \mathbb{R}^2 : x_2 = -x_1, 0 \leq x_1 \leq 1\}.$$

**2.** Calculate the polar of the set

$$\Omega = \{(x_1, x_2) \in \mathbb{R}^2 : (x_1 - 1)^2 + x_2^2 \leq 1\}.$$

**3.** Calculate the polar of the set

$$\Omega = \{(x_1, x_2) \in \mathbb{R}^2 : (x_1 - 1)^2 + x_2^2 = 1\}.$$

**4.** Prove part (e) of Proposition 5.5.

**5.** Prove the second assertion in Proposition 5.9.

**6.** Write the conditions in Example 5.13 in matrix form.

**7.** Prove that the Minkowski sum of two star convex domains is star convex.

**8.** Prove that the set-theoretic product of two star convex domains is star convex.

**9.** Show that the Minkowski sum of two ortho-convex sets is ortho-convex.

**10.** Prove that the intersection of ortho-convex sets is ortho-convex.

**11.** Calculate the infimal convolution of $f(x) = x$ and $g(x) = x^2$.

**12.** The Ace Novelty Company has determined that the profits are \$6, \$5, and \$4 for each type-A, type-B, and type-C souvenir that it plans to produce. To manufacture a type-A souvenir requires 2 minutes on machine I, 1 minute on machine II, and 2 minutes on machine III. A type-B souvenir requires 1 minute on Machine I, 3 minutes on machine II, and 1 minute on machine III. A type-C souvenir requires 1 minute on machine I and 2 minutes on each of machines II and III. There are 3 hours available on machine I, 5 hours available on machine II, and 4 hours available on machine III for manufacturing these souvenirs each day. How many souvenirs of each type should Ace Novelty make per day in order to maximize its profit?

**13.** A plant can manufacture five products $P_1, P_2, P_3, P_4$, and $P_5$. The plant consists of two work areas: the job shop area $A_1$ and the assembly area $A_2$. The time required to process one unit of product $P_j$ in work area $A_i$ is $p_{ij}$ (in hours), for $i = 1, 2$ and $j = 1, \ldots, 5$. The weekly capacity of work area $A_i$ is $C_i$ (in hours). The company can sell all it produces of product $P_j$ at a profit of $s_j$ , for $j = 1, \ldots, 5$.

The plant manager thought of writing a linear program to maximize profits, but never actually did for the following reason: From past experience, he observed that the plant operates best when at most two products are manufactured at a time. He believes that if he uses linear programming, the optimal solution will consist of producing all five products. Do you agree with him? Explain based on your knowledge of linear programming.

# Chapter 6

# The MiniMax Theorem

**Prologue:** The MiniMax theorem of John von Neumann has proved to be one of the truly influential ideas in optimization theory. von Neumann himself said that his studies in this area had produced nothing worth publishing until he discovered the MiniMax theorem.

Here we prove the theorem in some detail and then provide some illustrations of the ideas.

## 6.1 von Neumann's Theorem

We conclude the book with a treatment of John von Neumann's MiniMax theorem. Von Neumann originally conceived of this theorem in 1928 (see [VON]) in the context of game theory. He gave several proofs over the years, some of them relating in a profound way to Brouwer's fixed-point theorem. The argument that we present here, due to Jürgen Kindler [KIN], is close to von Neumann's original proof. What we treat here is actually Maurice Sion's version of the theorem, for which see [SIO]. We begin by stating the theorem. After the proof, we shall say a few words about applications.

THEOREM 6.1 *Let $X$ and $Y$ be nonempty, convex, compact subsets of $\mathbb{R}^N$. Let $f : X \times Y \to \mathbb{R}$ be a function that is continuous and concave in the first variable and continuous and convex in the second variable. Then*

$$\min_{y\in Y} \max_{x\in X} f(x,y) = \max_{x\in X} \min_{y\in Y} f(x,y) .$$

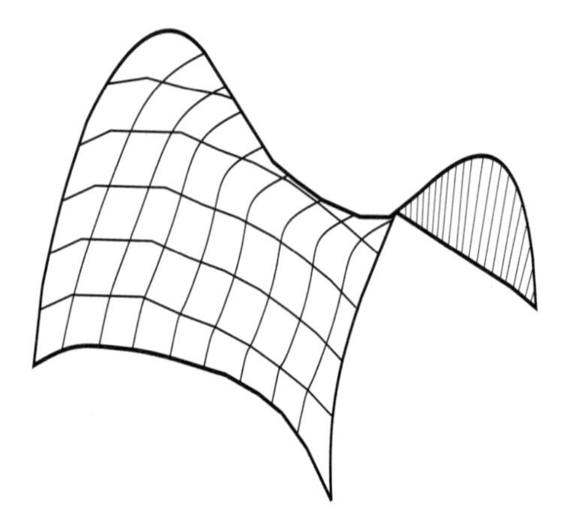

Figure 6.1: A saddle surface.

**Remark 6.2** In visualizing this result, the reader should think of a saddle point (as in basic calculus of several variables)—see Figure 6.1. Indeed, many proofs of the MiniMax theorem reduce the problem to proving the existence of a saddle point.

A further comment is that the hypotheses of the theorem guarantee that the set

$$U_\alpha \equiv \{x \in X : f(x,y) \geq \alpha\}$$

for fixed $y \in Y$ is closed and convex and also the set

$$L^\beta \equiv \{y \in Y : f(x,y) \leq \beta\}$$

for fixed $x \in X$ is closed and convex. It is not difficult to see that the maxima and minima recorded in the statement of the theorem exist in these circumstances.

**Proof of Theorem 6.1:** The argument hinges on the following lemma:

LEMMA 6.3 *Under the hypotheses of the theorem, let $K_1, K_2$ be nonempty, convex, compact subsets of $X$ with $X = K_1 \cup K_2$. Then*

$$\xi \equiv \max_{j\in\{1,2\}} \min_{y\in Y} \max_{x\in K_j} f(x,y) = \beta \equiv \min_{y\in Y} \max_{x\in X} f(x,y) .$$

**Proof:** Select $z_j \in Y$ with $\max_{x\in K_j} f(x, z_j) \leq \xi$ for $i = 1, 2$. Then certainly $\xi \geq \max_{x\in X} \min_{y\in I} f(x,y)$, where $I = [z_1, z_2]$ (an interval) is the convex hull of $\{z_1, z_2\}$. Suppose that $\min_{y\in I} \max_{x\in X} f(x,y) > \delta > \xi$ for some real

$\delta$. Then we obtain that $\cap_{x \in X} L^\delta(x) \cap I = \emptyset$ and $\cap_{y \in I} U_\delta(y) = \emptyset$. By the 1-dimensional Helly theorem, there thus exist $x_1, x_1 \in X$ with

$$L^\delta(x_1) \cap L^\delta(x_2) \cap I = \emptyset . \tag{6.3.1}$$

Also there exist $y_1, y_2 \in I$ with

$$U_\delta(y_1) \cap U_\delta(y_2) \cap \{x_1, x_2\} = \emptyset .$$

This all implies that

$$L^\delta(x_j) \cap \{y_1, y_2\} \neq \emptyset \quad \text{for } j = 1, 2 . \tag{6.3.2}$$

Since the sets $L^\delta(x_j)$ are closed and $[y_1, y_2]$ is connected, equations (6.3.1) and (6.3.2) imply the existence of a $y_0$ such that

$$y_0 \in [y_1, y_2] \setminus \left( L^\delta(x_1) \cup L^\delta(x_2) \right) . \tag{6.3.3}$$

A similar argument shows that there is an $x_0$ such that

$$x_0 \in [x_1, x_2] \setminus \left( U_\delta(y_1) \cup U_\delta(y_2) \right) l \tag{6.3.4}$$

By (6.3.3) we know that $[x_1, x_2] \subset U_\eta(y_0)$ for some $\eta > \delta$. Since $x_0 \in [x_1, x_2] \subset U_\eta(y_0)$, we conclude that $f(x_0, y_0) > \delta$. Similarly, (6.3.4) implies that $f(x_0, y_0) < \delta$. That is a contradiction.

Thus we have that $\xi \geq \min_{y \in I} \max_{x \in X} f(x, y) \geq \beta$ (which in turn is $\geq \xi$).

□

**Remark 6.4** The lemma is quite satisfying, both in its statement and its proof. For the statement shows the role of convex functions. And the proof demonstrates the role of convex sets (especially in the use of Helly's theorem).

**Proof of the Theorem:** Fix a finite subset $A$ of $X$ and set

$$\beta = \min_{y \in Y} \max_{x \in [A]} f(x, y) ,$$

where $[A]$ denotes the convex hull of $A$. Let $\mathcal{K}$ denote the collection of all nonempty, closed, convex subsets $K$ of $[A]$ with

$$\min_{y \in Y} \max_{x \in K} f(x, y) = \beta .$$

Let $\mathcal{C}$ be a chain in $\mathcal{K}$ with respect to the partial ordering of set inclusion. The fact that $U_\beta(y) \cap C \neq \emptyset$ for all $C \in \mathcal{C}$ and $y \in Y$ implies that $U_\beta(y) \cap \bigcap \mathcal{C} \neq \emptyset$ for each $y \in Y$ (i.e., $\cap \mathcal{C} \in \mathcal{K}$). As a result, $\mathcal{K}$ is inductively ordered by inclusion.

By Zorn's lemma, $\mathcal{K}$ has a minimal element $K_0$. Suppose now that $K_0$ has proper closed convex subsets $K_1$ and $K_2$ with $K_0 = K_1 \cup K_2$. Then the lemma tells us that either $K_1 \in \mathcal{K}$ or $K_2 \in \mathcal{K}$. This contradicts the minimality of $K_0$. Therefore $K_0$ must be a singleton $\{x_0\}$. This result implies that

$$\begin{aligned} \alpha &\equiv \max_{x \in X} \min_{y \in Y} f(x, y) \\ &\geq \min_{y \in Y} f(x_0, y) \\ &= \min_{y \in Y} \max_{x \in [A]} f(x, y) \\ &\geq \min_{y \in Y} \max_{x \in A} f(x, y) . \end{aligned}$$

As a result, $\cap_{x \in A} L^\alpha(x) \neq \emptyset$ for each finite subset $A$ of $X$.

Since $Y$ is compact and all the sets $L^\alpha(x)$ are closed, there exists a $y_0 \in \cap_{x \in X} L^\alpha(x)$. Therefore

$$\alpha \geq \max_{x \in X} f(x, y_0) \geq \min_{y \in Y} \max_{x \in X} f(x, y)$$

and this last is $\geq \alpha$. That completes the proof. □

Now why is the minimax theorem so important? Certainly the earliest applications of the idea were to two-person, zero-sum games. This is what von Neumann and Morgenstern were thinking about in [VNM]. A nice way to interpret the MiniMax theorem in terms of game theory is this:

For every two-person, zero-sum game with finitely many strategies, there exists a value $V$ and a mixed strategy for each player, such that

**(a)** Given Player 2's strategy, the best payoff possible for Player 1 is $V$, and

**(b)** Given Player 1's strategy, the best payoff possible for Player 2 is $-V$.

Equivalently, Player 1's strategy guarantees him a payoff of $V$ regardless of Player 2's strategy, and similarly Player 2 can guarantee himself a payoff of $-V$. The name minimax arises because each player minimizes the maximum payoff possible for the other—since the game is zero-sum, he also minimizes his own maximum loss (i.e., maximize his minimum payoff).

In more recent years this set of ideas has been generalized to be included in the theory of the Nash equilibrium. John Nash won the Nobel Prize in Economics for that work.

# Exercises

1. The following example of a zero-sum game, where A and B make simultaneous moves, illustrates minimax solutions. Suppose each player has three choices and consider the payoff matrix for A displayed at right. Assume the payoff matrix for B is the same matrix with the signs reversed (i.e., if the choices are A1 and B1 then B pays 3 to A). Then, the minimax choice for A is A2 since the worst possible result is then having to pay 1, while the simple minimax choice for B is B2 since the worst possible result is then no payment. However, this solution is not stable, since if B believes A will choose A2 then B will choose B1 to gain 1; then if A believes B will choose B1 then A will choose A1 to gain 3; and then B will choose B2; and eventually both players will realize the difficulty of making a choice. So a more stable strategy is needed. Devise such a strategy.

2. Suppose the game being played only has a maximum of two possible moves per player each turn. Because of the limitation of computation resources, the tree is limited to a look-ahead of 4 moves.

   The algorithm evaluates each leaf node using a heuristic evaluation function. The moves where the maximizing player wins are assigned with positive infinity, while the moves that lead to a win of the minimizing player are assigned with negative infinity. At level 3, the algorithm will choose, for each node, the smallest of the child node values, and assign it to that same node (e.g., the node on the left will choose the minimum between "10" and "+8," therefore assigning the value "10" to itself). The next step, in level 2, consists of choosing for each node the largest of the child node values. Once again, the values are assigned to each parent node. The algorithm continues evaluating the maximum and minimum values of the child nodes alternately until it reaches the root node, where it chooses the move with the largest value (represented in the figure with a blue arrow). This is the move that the player should make in order to minimize the maximum possible loss.

   Draw a diagram to illustrate the analysis discussed here.

# Chapter 7

# Concluding Remarks

We have attempted in this book to provide the analytic tools so that convexity can be used in works of geometric analysis. Basic among these tools is the effective use of convex functions, convex exhaustion functions, and convex defining functions.

As we have said earlier, convex functions are much more flexible and powerful tools than convex sets. Convex functions carry all the information about convex sets—and much more.

We hope that this point of view will prove useful, and provide the reader with motivation for further study.

# Appendix: Technical Tools

**hyperplane (Section 1.1)** A *hyperplane* in $\mathbb{R}^N$ is a linear subspace of dimension $(N-1)$. Typically a hyperplane is described by a linear equation:

$$a_1x_1 + a_2x_2 + \cdots + a_Nx_N = c\,.$$

The tangent space at a boundary point of a domain in $\mathbb{R}^N$ is a hyperplane.

**$C^{1,1}$ function (Section 1.3)** Let $f$ be a function with domain the interval $(a,b)$. We say that $f$ is $C^1$ if it is continously differentiable—that is, it is differentiable at all points and the derivative function is continuous. A function $g$ on $(a,b)$ is *Lipschitz* if $|g(x)-g(y)| \le C|x-y|$ for some constant $C$ and all $x, y \in (a,b)$. Finally, a function $h$ on $(a,b)$ is $C^{1,1}$ if $h$ is $C^1$ and the derivative function $h'$ is Lipschitz. Note that the property of being $C^{1,1}$ is close to, but just shy of, being $C^2$ or twice continuously differentiable.

**tubular neighborhood (Section 2.1)** Let $\partial\Omega$ be the boundary of a domain in $\mathbb{R}^N$ with $C^2$ boundary. A *tubular neighborhood* of $\partial\Omega$ is an open set $U \supset \partial\Omega$ such that each point $P$ of $U$ has a unique nearest point $\pi(P)$ in $\partial\Omega$. Of course $\pi(P)$ will simply be the orthogonal projection of $P$ into $\partial\Omega$.

We construct $U$ by using the inverse function theorem. To wit, consider the mapping

$$\Psi : \partial\Omega \times (-1,1) \to \mathbb{R}^N$$

given by

$$(x,t) \longmapsto x + t\nu_x\,,$$

where $\nu_x$ is the outward-pointing normal vector to $\partial\Omega$ at $x$.

At a point $(x_0, 0)$, with $x_0 \in \partial\Omega$, the Jacobian of this mapping will have nonvanishing determinant. So there will be a neighborhood $W_{x_0}$ of $(x_0, 0)$ so that the restriction of $\Psi$ to $W_{x_0}$ is invertible. But this just says that every point $z$ of $\Omega$ that is near to $x_0$ can be written uniquely as $z = y + t\nu_y$ for some $y \in \partial\Omega$. The union of all the neighborhoods $W_{x_0}$ over $x_0 \in \partial\Omega$ gives the tubular neighborhood $U$ that we seek.

**subharmonic function (Section 2.4)** A *subharmonic function* on a domain $\Omega \subseteq \mathbb{R}^N$ is defined to be a continuous function $u$ on $\Omega$ with the property that, whenever the ball $B(P, r) \subseteq \Omega$, then

$$u(P) \leq \frac{1}{|B(P,r)|} \int_{B(P,r)} u(x)\, dx_1 dx_2 \cdots dx_N \,.$$

Here we use absolute value signs to denote the volume of a set. It is known that subharmonic functions satisfy the maximum principle. That is to say, a subharmonic function never assumes a local maximum value in the interior of a domain—only at the boundary. A $C^2$ function $u$ is subharmonic if and only if $\triangle u \geq 0.1$

For further properties of subharmonic functions, refer to [KRA1, § 2.1].

**radial function (Section 2.4)** A function $f$ on $\mathbb{R}^N$ is *radial* if $f(x) = f(x')$ whenever $|x| = |x'|$. Such a function is *radial about the origin.*

Sometimes we want to allow the radial property about another point $a$. We say that $f$ is *radial about* $a$ if $f(x) = f(x')$ whenever $|x - a| = |x' - a|$.

**Sard's theorem (Section 2.4)** Let $f$ be a smooth function from $\mathbb{R}^N$ to $\mathbb{R}^M$. A point $x$ is a singular point of $f$ if the Jacobian matrix at $x$ has rank less than $M$. Then Sard says that the image of the set of singular points has measure zero. In particular, the image of the singular set has no interior.

**vanishing to order $k$ (Section 3.2)** Let $f$ be a function on an open set $U \subseteq \mathbb{R}^N$ and let $P \in \Omega$. We say that $f$ *vanishes to order* $k$ *at* $P$ if any derivative of $f$, up to and including order $k$, vanishes at $P$. Thus if $f(P) = 0$ but $\nabla f(P) \neq 0$ then we say that $f$ vanishes to order 0. If $f(P) = 0$, $\nabla f(P) = 0$, $\nabla^2 f(P) = 0$, and $\nabla^3 f(P) \neq 0$, then we say that $f$ vanishes to order 2. The function $f(x) = x^2$ vanishes to order 1 at the origin. More generally, the function $g(x) = x^{k+1}$ vanishes to order $k$ at the origin. In several variables, the function $w(x) = |x|^{k+1}$ vanishes to order $k$ at the origin, while the function $v(x) = (x_1^2 + x_2^4)^m$ vanishes to order $2m - 1$ at the origin.

**orthogonal mapping (Section 3.2)** A mapping $\Phi : \mathbb{R}^N \to \mathbb{R}^N$ is said to be *orthogonal* if it preserves the Euclidean inner product. In particular, it respects orthogonality and preserves length.

An orthogonal mapping is given by a matrix with the properties that

**(a)** Each row is a unit vector;

**(b)** The rows are orthogonal to each other;

**(c)** The matrix has determiant 1.

**affine function (Section 3.5)** A function $\Phi : \mathbb{R}^N \to \mathbb{R}^M$ is *affine* if there is an element $\alpha \in \mathbb{R}^M$ so that the map $\Psi(x) = \Phi(x) - \alpha$ is linear. Thus an affine mapping is the translate of a linear mapping. For example, the map

$$\Phi(x_1, x_2) = (3x_2 - 2, 4x_1 + 3)$$

is affine (but not linear). Speaking synthetically, an affine mapping is one that preserves points, lines, and planes.

**Hausdorff distance (Section 3.6)** The Hausdorff distance is a metric on the collection of compact sets. In detail, if $A$ and $B$ are compact sets, then

$$d_{\mathcal{H}}(A, B) = \max\{\sup_{a\in A} \inf_{b\in B} |a - b|\, , \, \sup_{b\in B} \inf_{a\in A} |a - b|\}\, .$$

The Hausdorff distance between two disjoint compact sets is the distance between the nearest points. The Hausdorff distance between the interval $[0, 2]$ and the interval $[1, 3]$ is 1.

**seminorm (Section 5.1)** A *seminorm* on $\mathbb{R}^N$ is a function $\rho : \mathbb{R}^N \to \mathbb{R}^+$ with the properties

**(a)** $\rho(ax) = |a|\rho(x)$ for any $a \in \mathbb{R}$ and any $x \in \mathbb{R}^N$;

**(b)** $\rho(x + y) \leq \rho(x) + \rho(y)$ for any $x, y \in \mathbb{R}^N$.

A *norm* would have the additional property that $\rho(x) = 0$ implies that $x = 0$.

An example of a seminorm on $\mathbb{R}^3$ is

$$\rho(x_1, x_2, x_3) = |x_1 + 2x_2 + x_3|\, .$$

This particular $\rho$ clearly satisfies **(a)** and **(b)** above. But note that $\rho(-1, 1, -1) = 0$.

# TABLE OF NOTATION

| Notation | Section | Definition |
|---|---|---|
| $\overline{PQ}$ | 1.0 | segment connecting $P$ to $Q$ |
| $\Omega$ | 1.0 | a domain |
| $\mathcal{H}$ | 1.2 | a hyperplane |
| $S^o$ | 1.3 | the polar of the set $S$ |
| $S^{oo}$ | 1.3 | the bipolar of the set $S$ |
| $A$ | 1.3 | a cone |
| $\partial\Omega$ | 2.1 | the boundary of $\Omega$ |
| $\rho$ | 2.1 | a defining function |
| $C^k$ | 2.1 | $k$ times continuously differentiable |
| $C^\infty$ | 2.1 | infinitely differentiable |
| $\nabla\rho$ | 2.1 | the gradient of $\rho$ |
| $U$ | 2.1 | a tubular neighborhood |
| $\varphi$ | 2.1 | a $C_c^\infty$ function |
| $\perp$ | 2.1 | perpendicular to |
| $\nu = \nu_P$ | 2.1 | the unit outward pointing normal vector at $P$ |
| $T_P(\partial\Omega)$ | 2.1 | tangent hyperplane to $\partial\Omega$ at $P$ |
| $\mathcal{P}$ | 2.1 | a hyperplane |
| $\mathcal{R}$ | 2.1 | remainder term in the Taylor expansion |
| $\left(\frac{\partial^2\rho}{\partial x_j \partial x_k}(P)\right)_{j,k=1}^N$ | 2.2 | the Hessian |
| $\rho_\lambda$ | 2.2 | a strongly convexified defining function |
| $p$ | 2.2 | the Minkowski functional |
| $\\| \ \\|$ | 2.3 | a norm |
| $\lambda$ | 2.4 | an exhaustion function |

| Notation | Section | Definition |
|---|---|---|
| $\widehat{K}_{\mathcal{F}}$ | 3.1 | the convex hull of $K$ with respect to $\mathcal{F}$ |
| $\{I_j\}$ | 3.1 | a collection of closed segments |
| $E_{2k}$ | 3.1 | $\{x_1^2 + x_2^{2k} < \infty\}$ |
| $E_{2\infty}$ | 3.1 | $\{x_1^2 + e^{-1/\lvert x_2\rvert^2} < \infty\}$ |
| $\mathcal{L}$ | 3.3 | an invertible linear mapping |
| $L$ | 3.4 | a support function |
| $\mathcal{E}(f)$ | 3.5 | the epigraph of $f$ |
| $\widehat{\Omega}$ | 3.6 | the bump of $\Omega$ |
| $X^*$ | 4.1 | the dual of $X$ |
| $\mathcal{P}$ | 4.1 | the collection of all compact extreme sets of $K$ |
| $A + B$ | 4.2 | the Minkowski sum of $A$ and $B$ |
| $\lvert A\rvert$ | 4.3 | the Lebesgue outer measure of $A$ |
| $S^o$ | 5.1 | the polar of $S$ |
| $S^{oo}$ | 5.1 | the bipolar of $S$ |
| $f'_r$ | 5.4 | the right derivative of $f$ |
| $f'_\ell$ | 5.4 | the left derivative of $f$ |
| $\Gamma(z)$ | 5.5 | the Gamma function |
| $f \sqcap g$ | 5.7 | the infimal convolution |

# GLOSSARY

**absolute convex hull** The absolute convex hull of a set $S$ is the intersection of all absolutely convex sets that contain $S$.

**absolutely convex set** A set $S$ is absolutely convex if, whenever $x_1, x_2 \in S$ and $|\lambda_1| + |\lambda_2| \leq 1$, then

$$\lambda_1 x_1 + \lambda_2 x_2 \in S\,.$$

**affine function** A function $L : \mathbb{R}^N \to \mathbb{R}$ that is the sum of a linear function and a constant function.

**analytically (weakly) convex boundary point** Let $\Omega \subseteq \mathbb{R}^N$ be a domain with $C^2$ boundary and $\rho$ a defining function for $\Omega$. Fix a point $P \in \partial\Omega$. We say that $\partial\Omega$ is analytically (weakly) *convex* at $P$ if

$$\sum_{j,k=1}^{N} \frac{\partial^2 \rho}{\partial x_j \partial x_k}(P) w_j w_k \geq 0, \quad \forall w \in T_P(\partial\Omega).$$

**analytically strongly convex boundary point** We say that $\partial\Omega$ is analytically strongly (strictly) convex at $P$ if the inequality

$$\sum_{j,k=1}^{N} \frac{\partial^2 \rho}{\partial x_j \partial x_k}(P) w_j w_k \geq 0, \quad \forall w \in T_P(\partial\Omega)$$

is strict whenever $w \neq 0$.

**augmented form** The form of a linear programming problem in which extra (slack) variables are introduced in order to replace all inequalities by equalities.

**balanced set** The set $S$ is balanced if, whenever $s \in S$, then $-s \in S$.

**bipolar** The bipolar of a set $S$ is the polar of the polar of $S$.

**Brunn-Minkowski inequality** An inequality that relates the measure of the Minkowski sum of two sets to the measures of the individual sets.

**bumping** Let $\Omega$ be a convex domain and $P \in \partial\Omega$. We say that $\widehat{\Omega}$ is a bumped domain if $\partial\widehat{\Omega}$ and $\partial\Omega$ coincide away from $P$ and $\widehat{\Omega}$ contains $P$ in its interior.

**$C^k$ boundary for a domain** A domain $\Omega$ has $C^k$ boundary if it has $C^k$ defining function.

**closed convex hull** The closed convex hull of a set $S$ is the intersection of all closed, convex sets that contain $S$.

**closed segment** The closed segment determined by points $P$ and $Q$ is the set

$$\overline{PQ} = \{(1-t)P + tQ : 0 \leq t \leq 1\}.$$

**cone** A set $A \subseteq \mathbb{R}^N$ is a cone if, whenever $a \in A$ and $\lambda \geq 0$, then $\lambda a \in A$.

**convex function** Let $F : \mathbb{R}^N \to \mathbb{R}$ be a function. We say that $F$ is *convex* if, for any $P, Q \in \mathbb{R}^N$ and any $0 \leq t \leq 1$, it holds that

$$F((1-t)P + tQ) \leq (1-t)F(P) + tF(Q).$$

**convex hull** The convex hull of a set $S$ is the intersection of all convex sets that contain $S$.

**convex hull with respect to a family of functions** Let $\Omega \subseteq \mathbb{R}^N$ be a domain and let $\mathcal{F}$ be a family of real-valued functions on $\Omega$. Let $K$ be a compact subset of $\Omega$. Then the *convex hull of $K$ in $\Omega$ with respect to $\mathcal{F}$* is defined to be

$$\widehat{K}_{\mathcal{F}} \equiv \left\{ x \in \Omega : f(x) \leq \sup_{t \in K} f(t) \text{ for all } f \in \mathcal{F} \right\}.$$

**convex of order $k$** Let $\Omega \subseteq \mathbb{R}^N$ be a domain and $P \in \partial\Omega$ a point at which the boundary is at least $C^k$ for $k$ a positive integer. We say that $P$ is *analytically convex of order* $k$ if

- The point $P$ is convex;
- The tangent plane to $\partial\Omega$ at $P$ has order of contact $k$ with the boundary at $P$.

A domain is convex of order $k$ if each boundary point is convex of order at most $k$, and some boundary point is convex of order exactly $k$.

**convex set** A domain $\Omega$ such that, if $P, Q \in \Omega$, then the closed segment $\overline{PQ} \subseteq \Omega$.

**curve** A continuous mapping either from $[0,1]$ into $\mathbb{R}^N$ or from the circle $S^1$ into $\mathbb{R}^N$.

**defining function** Let $\Omega \subseteq \mathbb{R}^N$ be a domain with $C^1$ boundary. A $C^1$ function $\rho : \mathbb{R}^N \to \mathbb{R}$ is called a *defining function* for $\Omega$ if

**1.** $\Omega = \{x \in \mathbb{R}^N : \rho(x) < 0\}$;

**2.** ${}^c\overline{\Omega} = \{x \in \mathbb{R}^N : \rho(x) > 0\}$;

**3.** $\nabla\rho(x) \neq 0 \ \ \forall x \in \partial\Omega$.

**domain** A connected, open set.

**domain convex with respect to a family of functions** We say that $\Omega$ is *convex* with respect to $\mathcal{F}$ provided $\widehat{K}_{\mathcal{F}}$ is compact in $\Omega$ whenever $K$ is. When the functions in $\mathcal{F}$ are complex-valued then $|f|$ replaces $f$ in the definition of $\widehat{K}_{\mathcal{F}}$.

**exhaustion function** Let $\Omega \subseteq \mathbb{R}^N$ be a bounded domain. We call a function

$$\lambda : \Omega \to \mathbb{R}$$

an *exhaustion function* if, for each $c \in \mathbb{R}$, the set

$$\lambda^{-1}((-\infty, c]) = \{x \in \Omega : \lambda(x) \leq c\}$$

is a compact subset of $\Omega$.

**extreme point** A point $P \in \partial\Omega$ is extreme (for $\Omega$ convex) if, whenever $P = (1-\lambda)x + \lambda y$ and $0 \le \lambda \le 1$, $x, y \in \overline{\Omega}$, then $x = y = P$.

**feasible array** An array that is ready for application of the simplex algorithm.

**Gamma function** The function, invented by Leonhard Euler, given by

$$\Gamma(z) = \int_0^\infty e^{-t} t^{z-1}\, dt\,.$$

**gauge** If $K$ is a convex set, then define

$$\|x\| = \inf\{\lambda \in (0,\infty) : \lambda^{-1}x \in K\}\,.$$

We call $\| \ \ \|$ a *gauge*.

**Gaussian elimination** An algorithm for solving a system of linear equations.

**geometrically convex** Convex according to the definition of "convex set" provided in this Glossary.

**Hessian** See *real Hessian.*

**Lipschitz function** A function $f$ on $\mathbb{R}^N$ that satisfies a condition

$$|f(x) - f(y)| \le C|x - y|$$

for some constant $C > 0$.

**midpoint convex function** We say that $F : \mathbb{R}^n \to \mathbb{R}$ is *midpoint convex* if, for any $P, Q \in \mathbb{R}^N$,

$$F((1/2)P + (1/2)Q) \le (1/2)F(P) + (1/2)F(Q)\,. \qquad (*)$$

**MiniMax theorem** An idea from game theory that describes the geometric analysis of a saddle point.

**Minkowski functional** Let $K \subseteq \mathbb{R}^N$ be convex. For $x \in \mathbb{R}^N$, define

$$p(x) = \inf\{r > 0 : x \in rK\}.$$

Then $p$ is a Minkowski functional for $K$.

**Minkowski sum** If $A$ and $B$ are sets in $\mathbb{R}^N$, then their Minkowski sum is the set

$$A + B = \{a + b : a \in A, b \in B\}.$$

**normal vector** For $\Omega$ with $C^1$ boundary, having defining function $\rho$, we think of $\nu_P = \nu = \langle \partial\rho/\partial x_1(P), \ldots, \partial\rho/\partial x_N(P)\rangle$ as the outward-pointing normal vector to $\partial\Omega$ at $P$.

**ortho-convex set** An *orthogonally convex* set.

**orthogonally convex** A domain $\Omega$ is orthogonally convex if any segment parallel to one of the coordinate axes that connects two points of $\Omega$ lies entirely in $\Omega$.

**partition of unity** A collection of $C^\infty$ functions with compact support that sums to 1.

**pivot column** The column of the feasible array at which the pivoting algorithm begins.

**pivoting algorithm** An algorithm for solving a linear programming problem.

**polar of a set** The polar of a set $S \subseteq \mathbb{R}^N$ is the set

$$S^o = \{\alpha \in \mathbb{R}^N : \alpha \cdot x \leq 1 \text{ for all } x \in S\}.$$

**real Hessian** The quadratic form

$$\left(\frac{\partial^2 \rho}{\partial x_j \partial x_k}(P)\right)_{j,k=1}^N$$

is frequently called the real Hessian of the function $\rho$.

**Sard's theorem** A result that says that the set of singular values of a smooth function has measure zero.

**segment characterization of convexity** Whenever $\{I_j\}_{j=1}^{\infty}$ is a collection of closed segments in the domain $\Omega$ and $\{\partial I_j\}$ is relatively compact in $\Omega$, then so is $\{I_j\}$.

**separation of convex sets by a hyperplane** A hyperplane $\mathcal{P}$ separates two disjoint convex sets $A$ and $B$ if $A$ lies on one side of $\mathcal{P}$ and $B$ lies on the other side of $\mathcal{P}$.

**star convex set** A domain $\Omega$ with a distinguished point $X \in \Omega$ (called the *star point*) so that, if $p \in \Omega$, then segment $\overline{Xp}$ lies in $\Omega$.

**strongly analytically convex** The point $P$ in $\partial\Omega$ is strongly (strictly) analytically convex if the Hessian of the defining function at $P$ is positive definite on the tangent space.

**strongly convex function** Let $F : \mathbb{R}^N \to \mathbb{R}$ be a function. We say that $F$ is *convex* if, for any $P, Q \in \mathbb{R}^N$ and any $0 \le t \le 1$, it holds that

$$F((1-t)P + tQ) < (1-t)F(P) + tF(Q) .$$

**subharmonic function** A function with nonnegative Laplacian. Equivalently, a function that satisfies a sub-mean-value property.

**support function** Let $\Omega \subseteq \mathbb{R}^N$ be a bounded, convex domain and $P \in \partial\Omega$. Let $T_P(\partial\Omega)$ be the tangent hyperplane at $P$. If $L$ is a linear function that is negative on $\Omega$ and positive on the other side of $T_P(\Omega)$, then we call $L$ a support function.

**support hyperplane** If $\Omega$ is convex, then any point $P \in \partial\Omega$ will still have one (or many) hyperplanes $\mathcal{P}$ such that $\mathcal{P} \cap \overline{\Omega} = \{P\}$. We call such a hyperplane a support hyperplane for $\partial\Omega$ at $P$.

**tangent hyperplane** The union of all the tangent vectors to $\partial\Omega$ at a point $P \in \partial\Omega$.

**tangent plane** The union of all the tangent vectors to $\partial\Omega$ at a point $P \in \partial\Omega$.

**tangent vector** Let $\Omega \subseteq \mathbb{R}^N$ have $C^1$ boundary and let $\rho$ be a $C^1$ defining function. Let $P \in \partial\Omega$. An $N-$tuple $w = (w_1, \ldots, w_N)$ of real numbers is called a tangent vector to $\partial\Omega$ at $P$ if

$$\sum_{j=1}^{N} (\partial\rho/\partial x_j)(P) \cdot w_j = 0.$$

**tubular neighborhood** Let $\Omega$ be a domain and $\partial\Omega$ its boundary. A tubular neighborhood of $\partial\Omega$ is an open set $U \supseteq \partial\Omega$ such that, if $x \in U$, then $x$ has a unique nearest point in $\partial\Omega$. Thus the orthogonal projection from $U$ to $\partial\Omega$ is well defined.

**vanishes to order *k*** Let $f$ be a function on an open set $U \subseteq \mathbb{R}^N$ and let $P \in \Omega$. We say that $f$ *vanishes to order* $k$ *at* $P$ if any derivative of $f$, up to and including order $k$, vanishes at $P$.

**weakly analytically convex** The point $P$ in $\partial\Omega$ is weakly analytically convex if the Hessian of the defining function at $P$ is positive semidefinite on the tangent space.

# BIBLIOGRAPHY


**[AHR]** J. Ahrens, The simplex method,
`www.pstcc.edu/facstaff/jahrens/math1630/simplex.pdf`

**[ADSZ]** M. Avriel, W. Diewart, S. Schaible, and I. Zang, *Generalized Concavity*, Springer, 2013.

**[BHS]** G. Bharali and B. Stensønes, Plurisubharmonic polynomials and bumping, *Math. Z.* 261(2009), 39–63.

**[BOF]** T. Bonneson and W. Fenchel, *Theorie der konvexen Körper*, Springer-Verlag, Berlin, 1934.

**[BOV]** J. Borwein and J. Vanderwerff, *Convex Functions: Constructions, Characterizations, and Counterexamples*, Cambridge University Press, Cambridge, 2010.

**[CAT]** F. S. Cater, Constructing nowhere differentiable functions from convex functions, *Real Analysis Exchange* 28(2002/2003), 617–621.

**[EVG]** L. C. Evans and R. Gariepy, *Measure Theory and Fine Properties of Functions*, CRC Press, Boca Raton, FL, 1992.

**[FED]** H. Federer, *Geometric Measure Theory*, Springer-Verlag, New York, 1969.

**[FEN]** W. Fenchel, Convexity through the ages, *Convexity and its Applications*, Birkhäuser, Basel, 1983, 120–130.

**[HIR]** M. Hirsch, *Differential Topology*, Springer-Verlag, New York, 1976.

**[HOR]** L. Hörmander, *Notions of Convexity*, Birkhäuser Publishing, Boston, MA, 1994.

**[JEN]** J. L. W. V. Jensen, Om konvexe Funktioner og Uligheder imellem Middelværdier, *Nyt Tidskrift fur Mathematik B* 16(1905), 49–68.

**[KIN]** J. Kindler, A simple proof of Sion's minimax theorem, *American Mathematical Monthly* 112(2005), 356–358.

**[KIS1]** C. Kiselman, How smooth is the shadow of a convex body?, *J. London Math. Society* 33(1986), 101–109.

**[KIS2]** C. Kiselman, Smoothness of vector sums of plane convex sets, *Math. Scand.* 60(1987), 239–252.

**[KRA1]** S. G. Krantz, *Function Theory of Several Complex Variables*, 2nd ed., American Mathematical Society, Providence, RI, 2001.

**[KRP1]** S. G. Krantz and H. R. Parks, *The Geometry of Domains in Space*, Birkhäuser Publishing, Boston, MA, 1999.

**[KRP2]** S. G. Krantz and H. R. Parks, *The Implicit Function Theorem*, Birkhäuser, Boston, MA, 2002.

**[KRP3]** S. G. Krantz and H. R. Parks, Distance to $C^k$ manifolds, *Jour. Diff. Equations* 40(1981), 116–120.

**[KRP4]** S. G. Krantz and H. R. Parks, On the vector sum of two convex sets in space, *Canadian Journal of Mathematics* 43(1991), 347–355.

**[LAU]** N. Lauritzen, *Undergraduate Convexity: From Fourier and Motzkin to Kuhn and Tucker*, World Scientific, Singapore, 2013.

**[LAY]** S. R. Lay, *Convex Sets and Their Applications*, John Wiley and Sons, New York, 1982.

**[LEM]** L. Lempert, La metrique Kobayashi et las representation des domains sur la boule, *Bull. Soc. Math. France* 109(1981), 427-474.

**[ONE]** B. O'Neill, *Elementary Differential Geometry,* Academic Press, New York, 1966.

**[ROC]** R. T. Rockafellar, *Convex Analysis*, Princeton University Press, Princeton, NJ, 1970.

**[RUD]** W. Rudin, *Principles of Mathematical Analysis*, 3rd ed., McGraw-Hill, New York, 1976.

**[SHI]** M. Shimrat, Simple proof of a theorem of P. Kirchberger, *Pacific J. Math.* 5(1955), 361–362.

**[SIM]** B. Simon, *Convexity: An Analytic Viewpoint*, Cambridge University Press, New York, 2011.

**[SIO]** M. Sion, On general minimax theorems *Pac. Jour. Math.* 8(1958), 171–176.

**[STE]** E. M. Stein, *The Boundary Behavior of Holomorphic Functions of Several Complex Variables*, Princeton University Press, Princeton, NJ, 1972.

**[TUY]** H. Tuy, *Convex Analysis and Global Optimization*, Kluwer, Boston, MA, 1998.

**[VAL]** F. A. Valentine, *Convex Sets*, McGraw-Hill, New York, 1964.

**[VLA]** V. Vladimirov, *Methods of the Theory of Functions of Several Complex Variables*, MIT Press, Cambridge, 1966.

**[VON]** J. von Neumann, Zur Theorie der Gesellschaftsspiele, *Math. Annalen* 100(1928), 295-320.

**[VNM]** J. von Neumann and O. Morgenstern, *Theory of Games and Economic Behavior*, Princeton University Press, Princeton, NJ, 1944.

**[ZYG]** A. Zygmund, *Trigonometric Series*, Cambridge University Press, Cambridge, UK, 1968.

# Index